# The Magnetic Sense

## *The Hidden Compass Guiding Animals*

Team Trex

Published by NobleTrex Press



For permissions and other inquiries, write to:

P.O. Box 3132, Framingham, MA 01701, USA

# Contents

# Introduction

Imagine, for a moment, that you are locked in a windowless room deep within the concrete bowels of a basement in Frankfurt, Germany. It is autumn. Outside, the leaves are turning, and the crisp air signals a time of change. But inside, the air is stagnant, the light is artificial, and the walls are thick enough to silence the city above. You are completely cut off from the sun, the stars, the wind, and landmarks. If I were to spin you around in a swivel chair and ask you to point toward Africa, you would be utterly helpless. You would likely guess, and you would likely be wrong.

Now, imagine you are a European robin.

In the late 1960s, this was exactly the scenario created by Wolfgang and Roswitha Wiltschko, a husband-and-wife team who would go on to revolutionize our understanding of animal behavior. They placed restless robins in a funnel-shaped cage inside a steel chamber. These birds were in the grip of *Zugunruhe*-migratory restlessness. They hopped frantically, fluttering against the paper lining of the funnel, leaving scratch marks that served as a record of their intended direction.

Despite having no view of the sky and no visual cues whatsoever, the robins consistently hopped toward the southwest-toward Spain and North Africa, their ancestral wintering grounds. They knew where to go. But then, the Wiltschkos did something that seemed like magic. They turned a dial on a power supply, activating large electromagnetic coils wrapping the room. They didn't touch the birds; they didn't make a sound; they didn't change the light. They simply rotated the magnetic field inside the room.

Immediately, the birds changed direction. When the magnetic "north" was shifted, the birds shifted their hopping to match it.

This experiment confirmed something that had been whispered about, debated, and often ridiculed for a century: life on Earth is tethered to the planet by an invisible force. While we humans walk through the world oblivious to the magnetic field arching through our bodies, for the robin-and for the sea turtle, the lobster, the moth, and the mole rat-this field is as tangible as the ground beneath our feet or the light hitting our retinas.

Welcome to the world of the magnetic sense. It is the only sense we humans lack entirely, and because of that, it remains the most mysterious. We can imagine what it is like to have the sharp eyes of an eagle or the sensitive nose of a bloodhound because we see and smell, too. But we cannot truly fathom what it feels like to sense the spin of the planet's molten core while flying over the Mediterranean.

This book is an invitation to step outside our own sensory limitations. We are going to explore a biological superpower that borders on science fiction, investigating how evolution has turned the entire globe into a navigational grid, and how life has found a way to plug into it.

***

To understand why this matters, we have to appreciate the sheer hostility of the problem nature solved. Navigation is difficult. If you have ever been turned around in a dense forest or lost your bearings in a fog, you know the panic of spatial disorientation. Now, scale that up. Imagine you are a loggerhead sea turtle hatchling. You are the size of a dinner plate, swimming alone in the vast, featureless blue of the Atlantic Gyre. You must navigate a circular loop of nearly 9,000 miles to avoid freezing waters that would kill you. You have no GPS, no sextant, and no parent to guide you. Yet, you stay the course.

Or consider the monarch butterfly. The individual completing the migration to the fir forests of Mexico has never been there before. It is the great-great-grandchild of the butterfly that left Mexico the previous spring. How does a brain the size of a pinhead inherit a map?

For decades, the answer was "instinct," a convenient placeholder word we use when we don't actually know what is happening. But "instinct" is not an explanation; it's a black box. The reality, as we will see, is far more mechanical, more physical, and more elegant. It turns out that animals are not wandering aimlessly; they are reading a heavy stream of data.

Our journey begins not with biology, but with the planet itself. The Earth is, in essence, a colossal bar magnet. But it is a messy one. The field varies in intensity from the equator to the poles; it tilts at different angles; it has hills and valleys of magnetic strength caused by iron deposits in the crust. To us, these are abstract lines on a geological diagram. To a navigator, they are signposts. We will see how this planetary web creates a coordinate system-a latitude and longitude that can be read if you possess the right hardware.

But possessing the hardware is the great puzzle. This is where the detective story of the magnetic sense becomes truly gripping. If you want to study vision, you look for the eye. If you want to study hearing, you look for the ear. But if you dissect a pigeon or a trout looking for the "magnetic organ," you will find nothing obvious. There is no magnetic nose on the face of a bird.

This elusiveness has driven scientists to the brink of frustration for fifty years. The search for the magnetoreceptor is one of the greatest "needle in a haystack" hunts in the history of biology. As we delve into the microscopic world, we will encounter two rival theories that have divided the scientific community, both of which appear to be true in different contexts.

One theory sounds like it belongs in a steampunk novel: that animals have literal compass needles growing inside their bodies. We now know that evolution learned to manufacture magnetite-highly magnetic iron crystals-billions of years ago. We will meet the researchers who have found these biological magnets connected to the nervous systems of trout and birds, essentially hardwiring the Earth's field directly into the brain.

The other theory is even stranger. It suggests that the magnetic sense is a form of quantum biology. We will explore the mind-bending possibility that when a European robin looks north, it is not just feeling a pull, but potentially *seeing* the magnetic field as a visual overlay, superimposed on its vision like a fighter pilot's heads-up display. This mechanism relies on a protein called cryptochrome and a process known as the "radical pair mechanism," where the spin of subatomic particles is entangled and influenced by the weak magnetic field of the Earth. If this hypothesis holds-and the evidence is mounting-it means that birds are mastering quantum mechanics in a warm, wet, messy biological eye, a feat that our best quantum computers struggle to replicate in a vacuum at absolute zero.

***

As we move through these chapters, we will treat the animal kingdom as a diverse cast of characters, each with a specific navigational problem to solve. We will look at the "map" versus the "compass"-the critical distinction between knowing which way you are facing and knowing where you actually are.

We will examine the Bogong moth, an insect that flies hundreds of miles at night to locate a specific cave in the Australian Alps, likely using a combination of magnetic cues and star patterns. We will look at the spiny lobster, which trudges across the seafloor in darkness, relying on a magnetic map so precise it can correct its course after being displaced into unfamiliar waters.

However, this is not just a book about how things work when they go right. It is also about what happens when they go wrong. We live in an era of unprecedented electromagnetic noise. Our cities glow with light, which interferes with the quantum compass, and our airwaves are thick with radio frequencies. We will explore the modern phenomenon of "electrosmog" and how it might be blinding migratory birds or confusing bees. As we alter the electromagnetic silence of the natural world, we may be severing the invisible threads that hold ecosystems together.

There is a concept in biology introduced by the Baltic-German biologist Jakob von Uexküll called the *Umwelt*. It refers to the self-centered world of an organism-the sensory bubble in which it lives. The tick waits for the scent of butyric acid; the bat lives in a world of rebounding sound waves. Humans, for the most part, live in an Umwelt of sight and sound. We are visual creatures, obsessed with what we can see.

The magnetic sense challenges our human arrogance. It forces us to admit that there are layers of reality right in front of our faces-literally passing through our bodies-that we are completely numb to. It reminds us that our perception of the Earth is incomplete.

When I look at a flock of geese cutting a V-formation across a gray November sky, I used to see a simple biological instinct. Now, I see a squadron of pilots reading a complex, shimmering instrument panel that spans the horizon. I see biology leveraging geophysics. I see a connection between the rock of the crust and the protein in the retina.

The goal of this book is to make that invisible world visible to you. We are going to strip away the surface of the planet and reveal the vibrant, oscillating grid that lies beneath. We are going to look into the eyes of birds and the brains of bees to see how nature engineered a sensor for a force that passes through stone.

The compass is hidden, but the needle is moving. Let's find out how.

Chapter 1

# The Invisible Web Enveloping the World

*We live on the surface of a giant magnet, yet we remain oblivious to the lines of force arching over our heads and diving through our bodies. For the animal kingdom, however, this planetary shield acts as a ubiquitous, silent roadmap, providing coordinates where no landmarks exist. Unlocking the secrets of animal navigation requires first mapping the chaotic, shifting terrain of the geomagnetic field itself.*

## 1.1 The Geodynamo and the Great Dipole

A migrating bird does not need to understand geology to profit from it, but the planet does something quietly extraordinary beneath its feet. Earth behaves, most of the time, as if it carries a giant bar magnet inside it: one end pulls compass needles northward, invisible lines loop through the air and space, and a protective magnetic cocoon softens the solar wind. Yet there is no permanent magnet buried in the mantle, no hidden iron spear aligned with the axis of rotation. The source is more unsettling and more alive: motion.

The magnetic field is made by a generator the size of a planet, operating in darkness, sealed off from us by nearly three thousand kilometres of solid rock. The word *geodynamo* names the basic idea: a self-sustaining dynamo inside Earth, driven by fluid motion in an electrically conducting layer. The setting is the outer core, a deep global ocean of molten iron alloy that surrounds a solid inner core. It is hot enough for metal to flow, dense enough to carry electric current, and restless enough to keep stirring for billions of years.

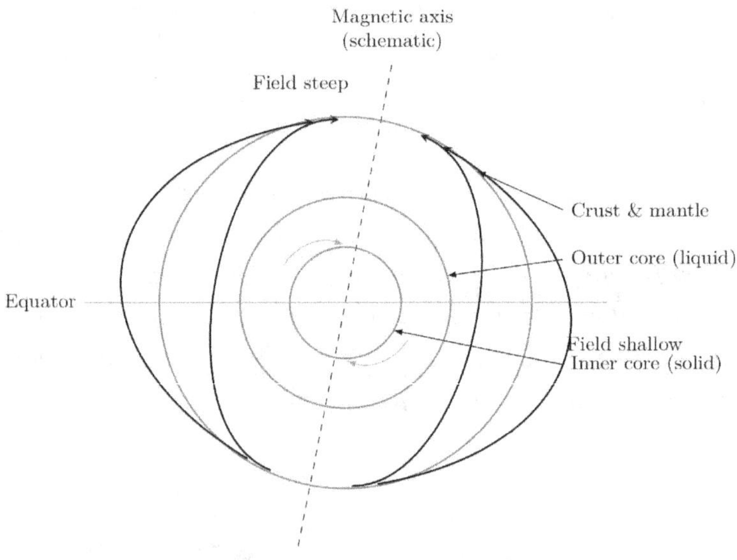


The image above gives the geometric vocabulary that matters for animal navigation. Magnetic field lines are not physical threads, but a way of drawing the direction a tiny compass needle would point at each location. Where the lines are steep, a compass needle wants to tilt downward into the ground; where the lines are shallow, it lies more nearly flat. At high latitudes the lines meet the surface almost perpendicularly. Near the equator they skim along nearly parallel. Those angles, and the strength of the field, become cues an animal can in principle measure. Before any biology enters the story, though, the field has to exist at all.

The leap from "Earth is magnetic" to "Earth is a dynamo" was not immediate. Early thinkers could be forgiven for imagining a permanent magnet: lodestones were known, and iron can hold magnetism. A particularly vivid step came from the English physician William Gilbert, who in 1600 argued that Earth itself behaves like a magnet. Gilbert's experiments, using a small magnetised sphere he called a *terrella*, reproduced the way a compass points and dips. It was a beautiful physical analogy: a globe that made needles behave as if they were standing on a planet. The analogy still serves, with one crucial correction. Earth behaves *like* a bar magnet in its large-scale geometry, but it is not magnetised in the way a piece of steel is magnetised. At core temperatures, any permanent magnetism would be destroyed. The field must be continually generated.

To see how, start with something familiar: an electric generator. In a power station, coils of wire rotate in a magnetic field, or magnets rotate past coils. Motion and magnetism conspire to create an electric current. The geodynamo runs this logic in a loop. In the outer core, there is already conducting fluid, and there is already some magnetic field-perhaps a weak "seed" field left over from the early Earth or generated by small instabilities. As molten iron moves through that field, it induces electric currents. Those currents create magnetic fields of their own. If the flow patterns are favourable, the new magnetic field reinforces the old one, and the whole system sustains itself. This is what "self-excited" means in dynamo theory: the machine makes the field that allows it to keep making the field.

That description is tidy; the reality is turbulent and, in a sense, precarious. The outer core is not a simple spinning drum. It is a rotating spherical shell, heated from below and cooled from above, with additional buoyancy released as the inner core grows and excludes lighter elements. The fluid is forced to move by convection: warm, buoyant material rises; cooler, denser material sinks. But because Earth rotates, those rising and sinking motions do not travel in straight lines. The Coriolis effect, the same influence that shapes large-scale weather patterns, twists the flow into columns and spirals aligned roughly with the rotation axis. The result is not random boiling. It is organised turbulence, biased into structures that are particularly good at winding magnetic field lines.

A useful mental picture is a stretchy bundle of elastic bands embedded in treacle. Stir the treacle, and the bands stretch, fold, and wrap around one another. A magnetic field behaves similarly. Where the fluid shears, field lines are stretched; where it rotates, they are twisted; where flows converge, they are concentrated. Stretching is important because it can amplify field strength, but amplification alone is not enough. A purely amplifying process would quickly run into a geometric dead end: the field would become tangled on small scales and cancel itself out. The geodynamo avoids that dead end because rotation tends to impose coherent patterns, repeatedly converting small-scale twists into a large-scale field. It is a complicated choreography, but it leaves a simple signature at the surface: a field that, when averaged over space and time, resembles a dipole.

A dipole is the field of a bar magnet: two opposed poles with field lines looping from one to the other. For Earth, the dipole axis is tilted relative to the rotation axis by about ten degrees, and it wanders slowly with time. In the ideal dipole, the field lines are symmetric and smooth, and the strength falls off with distance like $1/r^3$. That last fact matters far above our heads. A field that decays so quickly must be generated deep inside the planet; it cannot be produced by distant currents in space and still be as strong as it is at the surface. Space weather does add a shifting, sometimes violent magnetic contribution, but it sits on top of the main field rather than replacing it.

The word "pole" is where ordinary language starts to trip us. A compass needle's north-seeking end points toward the Arctic, so we call that direction "magnetic north." But a north-seeking pole is attracted to a magnetic *south* pole, in the same way opposite electric charges attract. So the region near geographic North is, in terms of magnetic polarity, more like a south pole. This is not pedantry. The direction of field lines depends on the choice: by convention, magnetic field lines emerge from a magnetic north pole and enter a magnetic south pole. For Earth's main field, that means the lines generally *enter* the planet in the Northern Hemisphere and *emerge* in the Southern Hemisphere. The diagram is drawn to match that convention: lines leaving the high southern latitudes and curving through space to dive back in near the north.

Those lines are not just directions; they contain three kinds of information that matter for navigation. First is *declination*: the horizontal angle between geographic north (true north, along a line of longitude) and magnetic north (the direction a compass points). Second is *inclination*, also called dip: the vertical angle the field makes with the local horizontal plane. Third is *intensity*: how strong the field is, often discussed either as a total strength or as vertical and horizontal components. Even if an animal never "knows" these as separate quantities, the physics makes them separable, and the geography makes them vary systematically.

The systematic part comes from the dipole geometry. At the magnetic equator-an imaginary line around the globe where the dipole field is exactly horizontal-the inclination is near zero. Move toward either magnetic pole and the field becomes steeper: inclination increases in magnitude until the lines are nearly vertical. That is why a compass needle, if it is free to tilt, does not sit level in London or New York. In the Northern Hemisphere it tends to dip down at the north end; in the Southern Hemisphere it dips down at the south end. Instrument makers learned long ago to "balance" needles differently for different latitudes, because the vertical component of Earth's field changes with where you are.

This vertical component is precisely the sort of cue a migrating animal could use without ever identifying a north end or a south end. A dipole has a built-in symmetry: the inclination flips sign across the magnetic equator. That means a sensor that can detect "field tilts downward" versus "field tilts upward" could distinguish hemispheres. In the 1970s, Wolfgang Wiltschko and Roswitha Wiltschko showed that European robins use an inclination compass: their orientation depends on the angle of the field lines, not on the polarity. Reverse the magnetic field (so north becomes south), but keep the inclination the same, and the birds behave as if nothing fundamental changed. That behavioural result makes immediate sense in the dipole picture. An inclination compass reads the geometry of the field lines, not the labels we stamp on the poles.

The dipole also creates a broad gradient in intensity. The field is generally stronger toward the poles and weaker toward the equator. This is not because the poles are "closer to the magnet," but because of how the field lines crowd. Where the lines are more vertical, they are more densely packed per unit area crossing the surface, and intensity is higher. In a perfect dipole, both inclination and intensity are predictable functions of latitude. Real Earth departs from perfection, but the dipole remains the dominant component, a kind of bass note under the planet's magnetic music.

If the dipole were the whole story, the magnetic sense would be almost too easy. A creature would carry, in effect, a built-in coordinate system: head toward the direction that keeps the field angle constant; seek the intensity value that matches home. But the same physics that produces the dipole also guarantees imperfection. The outer core is not a rigid rotor producing a pristine field; it is chaotic flow producing a field that changes. On top of the dipole sit higher-order components-quadrupoles and more complex shapes-that distort the lines. In some places, crustal rocks add their own frozen-in magnetism, locally tugging the field away from what the core alone would produce. And in the near-Earth environment, electric currents in the ionosphere and magnetosphere add transient perturbations. The dipole picture is the starting point because it supplies the global geometry; it is not the final map.

Even at the level of the core, there is an important emotional fact about the geodynamo: it is neither guaranteed nor gentle. Magnetic fields store energy, and moving a conducting fluid through a field can dissipate energy as heat through electrical resistance (Ohmic losses). The dynamo has to be powered continuously. Earth pays the bill through cooling-heat leaking out of the core into the mantle-and through compositional buoyancy as the inner core solidifies, releasing lighter elements into the liquid outer core and making them rise. One can think of the geodynamo as a by-product of a larger planetary drama: a hot planet losing heat into space. The same process that drives plate tectonics and volcanism, at different depths and through different mechanisms, also helps keep the magnetic field alive.

A historical anecdote brings home how physical this invisible structure became to human life. In 1831, during an era when global navigation depended on compasses and ships depended on navigation, James Clark Ross and his party located the North Magnetic Pole on the Boothia Peninsula in Arctic Canada. The "pole" they found was not a spike in the ground but a place where the compass needle, free to tilt, would point straight down. It was a geometric fact made tangible: a location where the field is vertical. For the explorers, it was an achievement of endurance and measurement. For the physics, it was evidence that the field has structure, that structure can be tracked, and that it is not permanently fixed. The magnetic pole Ross found is not at the same location today; it has drifted, sometimes rapidly, because the core flow that sustains the field also shifts it.

There is a final subtlety in the bar-magnet analogy worth keeping close, because it echoes in biology. A bar magnet has a fixed relationship between its internal magnetisation and its external field. Earth does not. The geodynamo is dynamic, and the field it produces is a compromise among competing flows, constrained by rotation, shaped by the boundary between core and mantle, and limited by the energy available. That means the field is not just a direction painted onto the world; it is an evolving physical environment. An animal that uses it inherits both its reliability and its restlessness.

Seen this way, the "great dipole" is less like a signpost and more like a climate: a global pattern with local weather. The large-scale lines arching from one hemisphere to the other give the planet a directional backbone, steep near the poles, shallow near the equator, strong where lines crowd and weaker where they spread. Out of molten metal, rotation, and time, an invisible geometry is continuously written onto the surface of Earth, and it is that geometry-simple enough to sketch, subtle enough to drift-that living creatures have learned to read.

## 1.2 Coordinates of the Invisible Map

A compass is an answer to a single, wonderfully simple question: *which way is north?* For a sailor in fog, that may be enough. For an animal trying to find a particular valley, beach, or breeding colony after travelling hundreds or thousands of kilometres, it is rarely enough. Knowing a direction is not the same as knowing a location. The difference is easy to feel in the body. Stand in a featureless desert with a compass and you can march north all day without learning whether you are nearer to home or farther from it. A direction gives an arrow; a map gives a place.

Earth's magnetic field offers both, because it is not merely a direction in the horizontal plane. It is a *vector field*: at every point it has a strength and a direction in three-dimensional space. That extra structure turns magnetism from a simple pointer into an information-rich environment. The trick is learning which pieces of information are stable enough, and patterned enough, to be useful.

The most widely used "magnetic coordinates" begin with three quantities that are measured routinely by geophysicists and used quietly by navigation systems. Each can be introduced with the same thought experiment: imagine a tiny compass needle that is free to move in all directions, not just spin on a pin.

The first quantity is *declination.* Put the needle on a level surface and it will settle pointing along the local magnetic field in the horizontal plane. Declination is the angle between that direction and true geographic north. It is what makes a magnetic compass slightly "wrong" in most places. In Boston the needle does not aim exactly at the geographic North Pole; in Sydney it does not aim exactly toward the meridian either. Declination is a reminder that the magnetic poles are not the geographic poles, and that the field is not a perfect dipole.

The second quantity is *inclination*, often called dip. Let the needle tilt. In the Northern Hemisphere, it tends to tip downward at its north-seeking end because the field lines enter the ground at an angle; in the Southern Hemisphere, the opposite end dips. Inclination is not a small correction. It can be steep enough that an unbalanced needle scrapes its housing. The planet's magnetism is three-dimensional, and inclination is the most direct way that fact makes itself felt.

The third quantity is *intensity*, the strength of the field. Intensity can be described as a total magnitude $F$ or broken into components: a vertical component $Z$, horizontal components $X$ (northward) and $Y$ (eastward), and a horizontal magnitude $H$. These letters can feel dry, but they matter because sensors-whether built by engineers or by evolution-often detect components, not abstract totals. A receptor aligned vertically will "feel" $Z$ more strongly than $H$; a horizontal compass system is sensitive to different combinations.

What makes these quantities map-like is that they vary across the globe in a patterned way. In a perfect dipole field, inclination changes smoothly with latitude: near the magnetic equator it is near $0°$ (field lines nearly parallel to the surface), and it grows in magnitude toward the poles (field lines becoming more vertical). Total intensity also tends to increase toward the poles as field lines crowd. Declination is trickier because it depends on how the magnetic axis is tilted and how the field departs from a simple dipole, but it too forms broad regions of similar values that can be drawn as contour lines.

A remarkable feature of this invisible geography is that it can provide two independent "axes" of information without ever referencing longitude and latitude explicitly. If inclination changes mainly with latitude, and intensity also changes mainly with latitude but not in exactly the same way, then the pair $(I, F)$ can act as a kind of bicoordinate signature. An animal that can sense both could, in principle, narrow its location far more than with a compass alone. It would not be a tidy grid of squares like a road atlas, but it would still be a map: certain combinations of inclination and intensity occur in some places and not others.

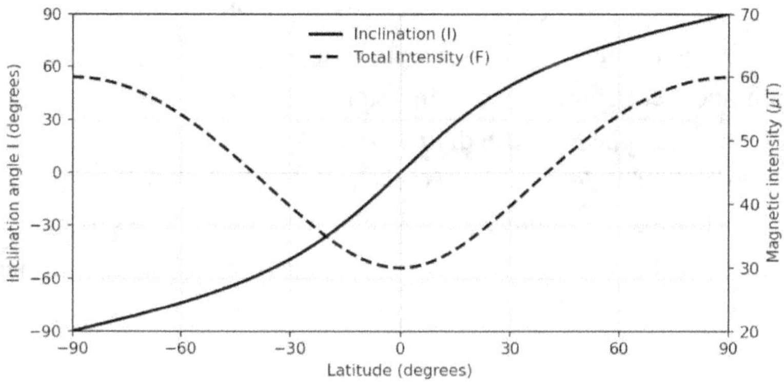


The plot is deliberately simple, like an idealised sketch in a notebook: the inclination curve rises steadily from the equator to the pole, and the intensity curve rises too, but with a different shape. Real Earth is messier. The magnetic equator is not the geographic equator; the curves shift with longitude; local anomalies wrinkle the smooth lines; and everything changes slowly with time. Yet the broad message survives: the magnetic field offers gradients that are large-scale and persistent enough to be informative.

Humans discovered the map-like nature of those gradients long before anyone suspected an animal might use them. In the late seventeenth century, Edmond Halley-better known for a comet-commanded voyages to measure declination across the Atlantic. He produced one of the first charts drawn not by coastlines but by an invisible quantity: lines connecting points of equal declination, now called isogonic lines. This was not just scientific curiosity. For navigation by compass, knowing local declination could be the difference between making landfall and missing it. Halley's chart turned magnetism into practical geography, even though the underlying physics of the geodynamo was centuries away.

Declination, inclination, and intensity are not only concepts; they are computed products in modern life. If a phone knows where it is and what time it is, software can estimate local magnetic declination and field components using standard global reference models. Two names appear repeatedly in the background of geophysics and navigation. The *International Geomagnetic Reference Field* (IGRF) is a scientific community standard: a mathematical description of Earth's main (core-generated) field, updated on a regular cadence. The *World Magnetic Model* (WMM) is an operational standard used widely in navigation and heading-reference systems. Both treat the main field as something that can be represented, to good approximation, by a series of global spatial patterns-spherical harmonics-whose coefficients change slowly with time. This is not because the field is simple, but because the large-scale part is coherent enough to be modelled, while the smallest-scale features and storm-time disturbances are not.

That modelling choice hints at an important biological point. An animal does not need the full richness of the field. It needs cues that are dependable at the scale of its movements. A migrating turtle crossing an ocean can profit from broad gradients in intensity and inclination; it cannot rely on a tiny crustal anomaly that extends only a few kilometres. In a sense, the animal's sensory world must include its own internal filtering: a way of treating the main field as signal and the more erratic components as noise.

How might the cues be used? Consider two very different tasks: *orientation* and *navigation*. Orientation means choosing and maintaining a direction: south in autumn for a songbird, offshore toward feeding grounds for a seal, toward the setting sun for a foraging insect. A magnetic compass-whether based on chemical processes such as cryptochrome radical pairs or on particles such as magnetite-can serve orientation by providing a stable reference axis. Navigation, in the stronger sense, means reaching a particular goal from unfamiliar territory. It implies some representation of location, however crude. Magnetic intensity and inclination are candidates for that representation precisely because they vary with place.

A concrete example comes from the open ocean, where landmarks are scarce and where magnetic cues are, if anything, more appealing because the sky can be overcast and waves can hide the shoreline. Sea turtles have long been a favourite system for studying magnetic navigation. Hatchlings entering the Atlantic must not only swim; they must avoid currents that would carry them into cold water, and later, as older juveniles, they roam over large areas yet show remarkable fidelity to broad regions of the ocean. Laboratory experiments have shown that changing the magnetic field around a turtle-altering inclination and intensity to values matching distant locations-can change its swimming direction in ways consistent with the idea that it interprets those fields as geographic information. For a young turtle, the field is not merely "north is that way"; it is more like "this combination of angle and strength belongs to a particular part of the Atlantic."

This sort of behaviour brings out a subtlety that often gets missed in casual descriptions of the magnetic sense. A magnetic compass and a magnetic map solve different problems and demand different kinds of stability. A compass can work even if the field strength varies, as long as direction is consistent and the sensor can maintain a reference. A map requires gradients that are not only detectable but also reliably associated with location, at least over the timescale relevant to the animal's memory. That is why secular variation-the slow drift of Earth's field over years and decades-matters for biology. An intensity value that once corresponded to "home" will gradually change. If an animal uses magnetic coordinates, it either needs to recalibrate, to rely on a moving target, or to combine magnetism with other cues.

The map-like use of magnetism also forces a confrontation with a geometric fact: magnetic coordinates are not the same as geographic coordinates. In geographic space, longitude lines meet at the poles and latitude lines are parallel circles. In magnetic space, lines of equal inclination and equal intensity do not form a perfect grid, and they can converge, diverge, and distort in ways that depend on regional field structure. In some parts of the world, two distant places can share similar inclination while differing in intensity; in other parts, intensity might change rapidly while inclination changes slowly. For an animal, this means that magnetic "addressing" may be more reliable in some regions than others.

Engineers working with near-Earth space physics face a related problem and have invented coordinate systems designed to respect magnetic field lines rather than geographic geometry. Systems such as corrected geomagnetic coordinates (CGM) and AACGM (altitude-adjusted corrected geomagnetic coordinates) exist because auroras, charged particles, and many ionospheric phenomena organise themselves along field lines. The motivation is practical: two points connected by the same magnetic field line can share physical behaviour even if their geographic coordinates differ. These systems come with caveats-there are regions where definitions become awkward or undefined-but the underlying impulse is revealing. When the field itself is the organising structure, it is natural to describe space in magnetic terms.

Animals, of course, do not publish coordinate definitions, and they do not need to. They only need the *functional* equivalents: sensory access to a directional axis (compass) and one or more gradients (map). That access can be surprisingly indirect. A receptor might not measure "inclination" as an angle. It might measure the relative activation of two pathways: one more sensitive to the vertical component, one more sensitive to the horizontal component. The brain could learn a ratio. Intensity might not be experienced as strength; it might be experienced as the degree of stimulation in a particular tissue, which can be compared over time. Nature often builds measurement devices by comparison rather than by absolute calibration.

Declination plays a different role. For human navigators, declination is a correction factor: turn your compass heading by a known number of degrees to aim toward true north. For animals, declination can in principle provide east-west information when used together with a celestial compass. If an animal has access to true north from stars or the Sun and magnetic north from its magnetic compass, the difference between them is declination. That difference varies geographically, and therefore could contribute to a map. There is experimental evidence in some systems that animals can learn relationships between magnetic and celestial cues, though how commonly declination is used as a map coordinate remains an open and fascinating question.

The deeper point is that "magnetic north" is only one projection of a richer field. Many popular accounts of magnetoreception implicitly imagine a built-in arrow pointing north like an internal compass app. That image is too thin. The field supplies an entire local signature: a direction in three dimensions, a steepness, a strength, and a skew relative to Earth's rotational geometry. The magnetic sense, whatever its physical mechanism, is a way of sampling that signature.

It is tempting to romanticise this sampling as flawless, as though an animal carries an infallible GPS. The reality, even in the idealised picture, is more like a set of contour lines on a weathered map. A sailor using Halley's isogonic lines would not expect a single declination value to identify a unique point on Earth; it identifies a curve of possibilities. Add a second quantity-inclination, or intensity-and the possibilities shrink, often to a small region where the curves intersect at distinctive angles. Add learning, memory, and integration with smell, vision, ocean currents, wind patterns, or landmarks, and the system becomes robust without needing to be perfect.

A final way to feel the power of magnetic coordinates is to imagine standing on a beach at night with all the familiar compasses taken away. No stars, no Moon, no distant lights. To you, the air gives no hint of north, and the sand gives no hint of latitude. Yet the field is there, crossing your body at a particular angle, pressing with a particular strength. If you could sense those two numbers the way you sense temperature and pressure, you would be standing inside a quiet, global pattern: a planet-sized gradient written into space. A creature that can read it does not merely carry a compass. It carries a way of belonging to the world's geometry, where "here" has a magnetic accent as distinctive as a coastline.

## 1.3 Anomalies, Storms, and Drifting Poles

A map only earns trust by staying put. The magnetic map does not. It changes slowly because the liquid core never stops moving, and it can change abruptly because Earth sits inside a stream of charged particles from the Sun. Even its spatial patterns are imperfect, tugged and wrinkled by iron-rich rocks in the crust. From the point of view of an animal trying to navigate, these are not academic complications. They are the difference between a reliable cue and a misleading one, between a compass that points steadily and a compass that jitters like a nervous hand.

Start with the quietest kind of change: the field's *secular variation*. "Secular" here means long-term, on the scale of years to centuries, not eternal. The main field is generated in the outer core by flows that shift as convection reorganises. Because the core is hidden, the clearest evidence for this slow evolution is written at the surface in numbers that have been recorded for centuries: declination at observatories, magnetic inclination, intensity, the wandering of the magnetic poles.

The North Magnetic Pole has been drifting since it was first located in 1831. That year matters because the "pole" was not a theoretical point but a destination reached by human bodies in cold air. James Clark Ross and his party, travelling by sledge and on foot on the Boothia Peninsula, measured a place where the field was essentially vertical: a compass needle allowed to tilt would point straight down. In the language of inclination, this was a spot near 90°. It was a triumph of persistence and instrumentation, and it fixed a headline fact: the planet's magnetic geometry could be pinned to the Earth's surface like a landmark.

But the landmark immediately began to move.

For much of the nineteenth and twentieth centuries, that motion was slow enough to be treated as a curiosity, a drifting point on a chart. By the late twentieth century it became something more: an operational problem. The pole's drift accelerated dramatically. In the 1990s and early 2000s it moved tens of kilometres per year, fast enough that global magnetic models used in navigation had to be updated more frequently and, in some cases, patched between scheduled releases. The explanation lies not in anything happening to the crust beneath the Arctic, but in the deep core field. Two broad "lobes" of magnetic flux at the core—mantle boundary, one under Canada and one under Siberia, compete to shape the northern high-latitude field. When the Canadian lobe weakens relative to the Siberian one, the pole is pulled across the Arctic toward Eurasia. A compass needle does not know any of this; it simply follows the local field direction. A migrating animal, likewise, inherits the outcome without access to the machinery.

The pole's drift is an arresting example because it is easy to picture, but it is only the most public face of secular variation. Declination at a given city changes over decades. Intensity trends upward or downward. Inclination slowly steepens or shallows. This matters because a magnetic "address" is, in effect, time-stamped. If a young animal imprints on a certain intensity value near its natal area and returns years later using that value as a cue, it must confront the fact that the field has changed during its absence. The problem is not hypothetical. Many long-lived species that migrate-sea turtles, salmon, seabirds-return after years. Their navigation systems are likely to depend on cues that are stable enough to be useful but flexible enough to be recalibrated.

There is a second kind of irregularity that is spatial rather than temporal: *magnetic anomalies*. The main field is generated in the core, but the crust contributes too. Some rocks acquire magnetisation when they form, as iron-bearing minerals align with the ambient field while cooling and then "freeze" in that orientation. In other places, chemical processes and later heating events modify the magnetisation. The result is a patchwork of local magnetic contributions superimposed on the large-scale field. To a human with a compass on a hike, these anomalies are usually minor; to a sensitive magnetoreceptor, they could be significant. A small hill with magnetised basalt can bend the local field enough to rotate a compass needle by a degree or two. For an animal that uses magnetism as one cue among several, such distortions may be manageable. For an animal relying heavily on magnetism at fine spatial scales, they are a potential source of error.

In the ocean, where rock exposures are distant and deep, one might expect anomalies to vanish. They do not. Seafloor spreading creates long stripes of alternating magnetic polarity recorded in basalt as it cools at mid-ocean ridges, a discovery that helped establish plate tectonics. Those stripes are far below the surface, and their influence at the sea surface is weaker than in direct rock contact, but they are not zero. More importantly, continental margins and volcanic seamounts can create broad anomalies that extend into the marine environment. The magnetic world is not simply a dipole plus small noise; it has regional personality.

One region has become famous not because compasses misbehave there, but because spacecraft do. Over the South Atlantic, roughly off the coast of Brazil and extending toward southern Africa, Earth's magnetic field is unusually weak. This is the South Atlantic Anomaly (SAA). It exists because the main field is not a perfect dipole and because certain reversed-flux features deep in the core reduce the field strength over that region. For satellites in low Earth orbit, a weaker field means less shielding from energetic particles trapped in the Van Allen radiation belts. Instruments can suffer glitches; detectors register spurious counts; operators sometimes switch sensitive systems off when passing through. The SAA is, in a very literal sense, a hole in Earth's magnetic umbrella.

For animal navigation, the SAA is intriguing for a different reason. A weak-field region changes intensity gradients and may introduce directional oddities. An animal using intensity as a coordinate could, in principle, experience the region as a magnetic "valley." The field there is still strong enough to be sensed, but the map-like contours are distorted. The SAA also changes with time, evolving in shape and even developing more complex structure. A biological system that uses magnetic intensity as a cue must be built for a world in which such large-scale features exist and drift.

Up to this point, the irregularities have been slow or geographically fixed. The third kind of disturbance is neither: *geomagnetic storms*. These are rapid, externally driven changes in the magnetic environment caused by conditions in space. Earth is embedded in the solar wind, a flow of charged particles and magnetic fields streaming outward from the Sun. When the solar wind is disturbed-by a coronal mass ejection or by fast streams interacting with slow wind-it can couple more strongly to Earth's magnetosphere. In simple terms, energy and momentum are transferred into the near-Earth magnetic system, driving electric currents in space and in the upper atmosphere. Those currents, in turn, modify the magnetic field measured on the ground.

Storms are not rare freak events. Mild and moderate ones occur regularly. Their signatures appear in indices that space physicists track: the $K$ index at individual observatories, $K_p$ as a global activity measure, and $D_{\mathrm{st}}$ (often written Dst) as a measure of the storm-time ring current that encircles Earth. The names are technical, but the phenomenon is tangible. During an active period, auroras can move to lower latitudes, radio communication can degrade, power grids can experience induced currents, and compass readings can wobble.

A nineteenth-century anecdote shows how unsettling such disturbances can be. In September 1859, during what is now called the Carrington event, a major solar storm produced brilliant auroras seen far from the polar regions and caused telegraph systems to misbehave. Operators reported sparks, shocks, and the ability to transmit briefly even after disconnecting power, because geomagnetically induced currents were flowing through the lines. Telegraph wires had unintentionally become antennas for space weather. For someone trying to navigate by compass during such a disturbance, the world would have felt slightly unmoored. The direction that had been dependable yesterday could drift by degrees today, and not because you moved, but because the field itself was being perturbed.

From the point of view of magnetoreception, storms pose an awkward question. If an animal's magnetic sense is exquisitely sensitive-as many experiments suggest-wouldn't it be thrown off by a storm? Sometimes, perhaps. Yet animals have evolved under the same Sun, with the same space weather. A plausible answer is that magnetic information is not used nakedly; it is weighted and cross-checked. When the magnetic field becomes noisy, other cues can take precedence: the pattern of stars, the Sun's arc, polarised light at twilight, wind direction, odours, landmarks, wave patterns. Another possibility is that the relevant sensory channels filter the field in time. A storm produces rapid fluctuations; the main field changes slowly. A nervous system that averages over minutes or hours would dampen high-frequency noise. A receptor that is tuned to long-term gradients might be inherently insensitive to fast jitter.

There is also a deeper physiological point. Some proposed magnetic sensing mechanisms, such as the radical pair processes associated with cryptochrome molecules, appear best suited to detecting the *direction* of the field rather than its absolute strength. Others, such as magnetite-based mechanisms, might be more sensitive to intensity. A storm perturbs both direction and intensity, but not necessarily in the same way at the same place and time. A multi-cue system could notice internal inconsistency: the compass says "north" has shifted while the intensity says "location" has not. In human terms, it would be like hearing two friends give slightly conflicting directions and deciding to trust the one who is usually right in that neighbourhood.

Irregularities also differ by latitude. High-latitude regions are more directly connected to space through magnetic field lines that funnel charged particles toward the atmosphere, which is why auroras occur there. The very places where inclination becomes steep and intensity grows-magnetically informative regions-are also places where storm-time currents can be strong. That creates an uncomfortable overlap between signal and disturbance. Migratory birds that travel at high latitudes, for example, inhabit a magnetic environment that is both geometrically clear and sometimes dynamically unruly.

Even without storms, there is a subtler kind of temporal change that animals must contend with: the daily rhythm of the ionosphere. Sunlight ionises the upper atmosphere during the day, altering current systems and producing regular diurnal variations in the ground magnetic field. These variations are typically small compared with the main field, but for a sensitive sensor they are real. Many experiments on magnetic orientation are therefore careful about time of day, local disturbances, and the stability of the experimental field, because biology can respond to differences that instruments might dismiss as minor.

All of this might make magnetic navigation sound fragile, as though it depends on a perfect field that Earth refuses to supply. Yet the opposite may be true. The magnetic environment has structure not in spite of its irregularities but partly because of them. A perfectly symmetrical dipole would provide clean latitude information but might leave more ambiguity in longitude. Regional distortions, if stable enough, can add uniqueness. A coastline is navigationally useful precisely because it breaks symmetry; so is a mountain range; so is a local smell. Magnetic anomalies and non-dipole structure can play a similar role, creating distinctive magnetic "landforms" that a learning animal could associate with particular regions.

The cost is that the landforms move. Secular variation drifts them over years. The North Magnetic Pole's rapid motion is a dramatic instance, but the same process shifts isolines of declination and intensity across continents. In practical navigation systems, this is handled by regular model updates: the field is treated as something that must be re-estimated. In biology, the updating must be done by experience. A young animal may calibrate its magnetic sense against other cues during development. An adult may refine its internal map through repeated journeys. A population may evolve strategies that tolerate error margins rather than demanding precision.

A final image helps: picture the magnetic field not as a rigid grid laid over Earth, but as a taut fabric that is constantly being tugged from below and fluttered from above. The tugs from below are slow, driven by core flow, shifting the position of poles and the shape of global contours. The flutters from above are fast, driven by solar wind, briefly rippling the fabric with storms. Local stitching-the crust's magnetisation-adds wrinkles that stay in place even as the larger fabric moves. An animal navigating through this world is not reading a printed map. It is reading a living one, learning which patterns persist, which patterns drift, and which patterns are momentary gusts.

When a bird lifts into night air or a turtle enters an open current, the magnetic field it senses is never exactly the same as the one it sensed last year, or even yesterday. Yet it is recognisably the same kind of field: a world-scale vector pattern with dependable gradients, regional quirks, and occasional tempests. The achievement is not that magnetism provides perfect coordinates, but that life has found ways to make use of a guide that is invisible, immense, and always slightly in motion.

Chapter 2

# Maps, Compasses, and the Quest for Home

*Migration is not merely movement; it is a geometric miracle performed without GPS. We dissect the twin problems of heading and position, exploring how animals synthesize disparate environmental cues to construct an internal representation of space that guides them across oceans and continents.*

## 2.1 The Compass and the Coordinate

A homing pigeon released in unfamiliar countryside rises, circles, and then sets off with unnerving confidence. To a human watching from the roadside, it looks like a small miracle: the bird has chosen a direction, committed to it, and vanished over the treeline. Yet that clean moment hides two very different problems that any traveller must solve. One is the problem of *which way to go*—the ability to keep a steady heading. The other is the problem of *where you are*—the ability to judge your position relative to a goal and adjust when you are off course. The first is a compass. The second is a map.

The distinction is easy to miss because in everyday language a compass sounds like navigation itself. But a compass is only a rule for maintaining a direction: "fly northwest," "swim shoreward," "head into the wind." A compass can be perfect and still be useless if the chosen direction is wrong. A map, in the biological sense, does not have to resemble a paper atlas with coastlines and labels. It is any internal information that can answer a brutally practical question: *what direction should I take from here to get closer to home?* If an animal can do that after being carried to an unfamiliar place—without retracing the route it came by—it has something map-like.

This way of thinking has a long pedigree in behavioural biology, and it became especially influential through the "map-and-compass" framework developed in the mid-twentieth century by Gustav Kramer. Kramer's insight was partly conceptual and partly experimental. Conceptually, he separated navigation into two components: a *compass* that provides a directional reference, and a *map* that provides position-related information. Experimentally, he and others helped popularize the displacement test: take an animal that knows where it wants to go, move it somewhere else without allowing it to monitor the journey, and see what it does next. A compass-only traveller behaves one way; a true navigator behaves another.

Imagine a bird that intends to migrate from point $A$ to point $B$, an inherited or learned seasonal goal. If released at $A$, even a simple compass is enough: choose the inherited migratory direction and hold it. But if the same bird is secretly transported to a new point $D$, far from its usual departure area, the logic changes. A bird with only a compass continues to fly as if it were still at $A$: it takes the same heading it would have taken anyway. It will not aim for $B$; it will aim along a line parallel to the intended route. A bird with map information does something more impressive: it compensates for displacement. It chooses a new heading from $D$ that points toward $B$, as if it can estimate where it is relative to the destination.

There is a particularly human way to appreciate the difference. Picture leaving a train station in a city you know well. You might not need a compass at all: you recognize the street, remember the café on the corner, and your feet take over. Now picture stepping off a train in an unfamiliar city in thick fog, with no landmarks and no GPS. If you have only a compass, you can walk north with great discipline, but you cannot know whether north is the direction that brings you closer to your hotel. What you need is at least one piece of map information—a clue about your position. "I'm west of the river," "I'm uphill from the old town," "I'm near the coast." With that, a compass becomes powerful, because it can convert position into action: *if I am west of the river, walk east; if I am south of the old town, walk north.*

In animals, the idea of a biological map can sound too cognitive, too close to deliberate cartography. It need not be. A map sense can be coarse, even categorical. It can be as simple as a decision rule triggered by gradients in the environment: *if the water tastes more like the estuary, turn seaward; if the smell is stronger to the left, go left.* In long-distance migrants, it may involve combining two weak pieces of information into something that behaves like coordinates. The key point is functional: after displacement, does the animal select a new bearing appropriate to the new location?

A classic historical anecdote makes the conceptual tension vivid. For centuries, sailors carried compasses across oceans long before they could reliably determine longitude. A magnetic compass could keep a ship on a steady heading, day after day. But without a secure sense of position—without a map solution—a ship could be perfectly oriented and still lost. The sea is unforgiving to travellers who confuse the ability to hold a direction with the ability to know where they are. The same is true for animals, except their instruments must be built from cells and nerves.

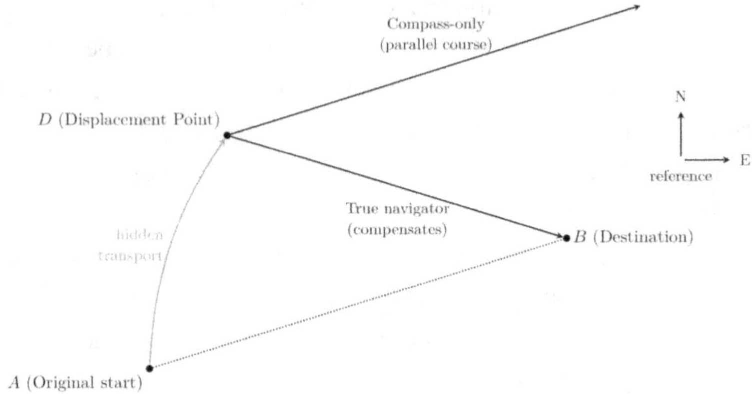


What makes this diagram scientifically useful is that it turns an inner sense into an observable signature. We cannot open an animal's head and watch its "map" light up. But we can infer map information if displacement produces a systematic change in chosen direction. In practice, the cleanest versions of such experiments use migrants at specific seasons (when motivation is strong and directional), and they try to prevent the animal from using simple route reversal. The animal is carried in darkness, or in a way that prevents it from tracking the journey using landmarks, sun position, or the pattern of stars.

Even with that care, the real world refuses to be a perfect laboratory. Animals are opportunists, and their navigation systems are layered. A bird might take a compensatory initial bearing and then switch to following rivers and coastlines. A sea turtle might use a map-like cue offshore and then rely on waves or smell nearer land. The map-and-compass distinction still helps, because it tells us which kind of information must exist at *some* stage: a heading reference alone cannot tell you that you have been moved.

The magnetic sense sits intriguingly at the boundary between these two functions. It is easy to see how a magnetic compass could work:

the Earth provides a global reference, and an animal needs only to extract direction from it. Many animals do precisely this. But magnetic information could, in principle, also contribute to a map, because the geomagnetic field varies across the planet in several measurable ways. Two features matter especially. The first is *inclination*: the angle at which field lines dip into the Earth. Near the magnetic equator, field lines are more horizontal; nearer the poles, they tilt steeply. The second is *total intensity*: the strength of the field, which generally increases toward the poles. Neither variable gives a street address, but together they can form a rough coordinate system, a pair of numbers whose combination differs by region.

A concrete example comes from sea turtles, which face a navigational problem that makes a mockery of land-based intuition. A juvenile turtle can spend years in the open ocean. There are no mountain silhouettes, no familiar intersections, no scent of home carried on the wind in any straightforward way. Yet turtles can reach broad target areas and, in some species, eventually return to natal regions. Laboratory studies have taken advantage of a simple but powerful trick. Instead of moving a turtle across the ocean, researchers can move the *magnetic field* around the turtle. With large coils, it is possible to generate magnetic conditions matching those found at different geographic locations. If the turtle responds differently to fields representing different places—for instance, orienting in one direction under the "northern" field and another direction under the "southern" field—that behaviour looks like the output of a map. It is as if the turtle interprets the magnetic field as a tag for region: *you are here*, therefore *go that way*.

The details matter. A compass response is typically consistent across locations: "swim east" might be the correct migratory heading regardless of where the animal is released (at least over some range). A map response, by contrast, should change with simulated location.

The same animal, under different magnetic combinations, should choose different headings that make geographical sense. This is exactly what has made marine migrants so compelling in the magnetic-map story. Kenneth and Catherine Lohmann's group, beginning in the early 2000s, helped establish this logic with sea turtles, showing that juveniles can respond to magnetic fields characteristic of distinct regions with orientation that would, in the ocean, tend to keep them within favourable currents and away from lethal dead-ends. It is difficult to read such results as a mere compass.

A parallel line of work has found similar signatures in juvenile Pacific salmon. Here the romance of the problem is matched by its brutality. Small fish entering the ocean for the first time face a moving, turbulent world, where a wrong turn can be fatal. Experiments by Nathan Putman and colleagues (2014) exposed naïve salmon to magnetic fields corresponding to different places along their likely oceanic routes and found that the fish oriented differently depending on the magnetic parameters. The interpretation is not that salmon carry a detailed ocean chart in their heads, but that they inherit a set of directional tendencies keyed to magnetic coordinates: in one magnetic "province" go one way; in another, go another. That is a map in the only sense that matters to survival.

Notice what has happened conceptually. The map-and-compass framework does not insist on an elaborate internal atlas. It asks whether the animal possesses information that changes the chosen compass bearing when the starting point changes. In turtles and salmon, the evidence suggests that magnetic parameters can play that map-like role, at least at regional scales. The magnetic sense can supply both the reference direction (compass) and, by exploiting gradients of inclination and intensity, a crude coordinate (map).

This raises an awkward but productive question: what prevents a magnetic map from being universally useful? The Earth's magnetic field is global, after all. The answer is that gradients can be steep in some places and shallow in others; they can be locally distorted by geology; and, most importantly, they change slowly over time. The field is not a painted grid on a stable globe but a restless, drifting phenomenon generated by motions in the planet's core. For an animal that lives a long time or returns decades later, the magnetic "address" of a place can shift. That does not make magnetic maps impossible; it makes them biological. A map that must function in a changing world can be learned, updated, or used as a broad region label rather than a precise coordinate.

The map-and-compass distinction also clarifies why the word *navigation* is used so loosely in popular accounts. Many experiments demonstrate *orientation*: an animal chooses a consistent direction relative to some cue. Orientation is already impressive, but it is not the whole story. True navigation, in the strict sense, implies that the direction is chosen appropriately for the animal's current position. A compass mechanism can produce beautiful, repeatable orientation in a lab cage and still tell us nothing about how an animal finds home from unfamiliar places.

Homing pigeons sit at the centre of this debate, historically and emotionally. Their feats were celebrated long before scientists could measure magnetic fields in a laboratory. Pigeons released far from their loft often depart in a homeward direction, suggesting a map component. Yet pigeons also use landmarks powerfully when near familiar terrain, and they respond to smells and winds in ways that implicate olfactory maps. The lesson is not that any one cue must explain everything, but that the logic of the problem forces at least two kinds of information. A pigeon that can find home from unfamiliar release sites must, at some stage, acquire position information. It may come from smell, from geomagnetism, from infrasound, from learned geography, or from a combination. But it cannot come from a compass alone.

That point can be sharpened by a thought experiment that mirrors the displacement diagram. Suppose an animal has a flawless magnetic compass and an inherited instruction: "fly southwest for 2,000 kilometres." From the correct departure area, this instruction can yield a successful migration. Now imagine the same animal blown 500 kilometres east by a storm at the start. If it continues southwest, it will run parallel to its intended track and may miss critical stopover sites or overshoot safe habitat. To correct, it needs to detect displacement. That requires at least one position-sensitive cue. If the magnetic field provides only a direction reference, it cannot signal that you have been moved. If the magnetic field provides both direction and a regional signature, it can.

The beauty of Kramer's framing is that it makes these arguments without claiming to know the machinery. At the level of behaviour, a compass-only strategy and a map-and-compass strategy leave different fingerprints. This is a gift to experimentation, because it allows researchers to design tests that separate functions even when the sensory organ remains mysterious. It also prevents a common category error: assuming that because an animal responds to magnetic north, magnetism must be the whole explanation of its navigation.

In the end, the compass and the coordinate are less like competing theories and more like complementary necessities. A coordinate without a compass is information without action: knowing you are west of home does not move you unless you can translate west into an eastward heading. A compass without a coordinate is action without context: moving in a steady direction does not guarantee moving toward anything that matters. The animals that impress us most are those that can bind the two together, turning an invisible, planet-sized pattern into a decision made in a moment: *from here, go that way*.

## 2.2 The Redundant Navigator

A navigator built around a single sense is a tragedy waiting for the right kind of bad day. Clear skies vanish into overcast. Coastlines retreat under fog. A river floods and redraws its banks. Even the Earth's magnetic field, so steady on a schoolroom globe, is locally warped by iron-rich rock and human metalwork. Animals that make a living by going somewhere specific have to treat the world as unreliable, and their own senses as fallible. The most striking feature of real navigation is not any one magical cue, but the habit of *not betting everything on one*.

Redundancy is an engineer's word, but it fits biology. A passenger plane has multiple instruments for altitude, multiple hydraulic systems, backup batteries, and procedures for what happens when readings conflict. No one calls that indecision; it is how you fly safely through clouds. Animals do something similar, except the backups are not bolted on as spares. They are interwoven from the start: a sun compass that works when the sky is readable; a star compass that takes over at night; a magnetic compass that functions even when clouds erase the heavens; odours that offer position when direction is not the problem; landmarks that transform a risky guess into a confident approach as the destination nears. The picture that emerges is not a single compass needle hidden in flesh, but a committee of senses that can vote, veto, and recalibrate each other.

The previous distinction between *compass* and *map* becomes more useful here, not less. Many of the cues animals use are excellent compasses but poor maps. The sun gives a direction reference, but it does not tell you where you are. The stars do the same. Even a magnetic compass is, by itself, a heading reference. Position information often comes from different channels: smells, landmarks, wind patterns, coastlines, and in some species a magnetic map-like sensitivity to regional features of the field. Redundancy is partly about having several compasses, and partly about having different kinds of information altogether, so that one can correct the other.

Migratory birds provide the most vivid demonstrations because their compasses can be tested in controlled situations. One of the most influential devices in this area is the Emlen funnel, a deceptively simple conical cage lined with paper. A bird in migratory condition, placed inside at night, tends to hop in the direction it wants to travel; the scratches it leaves on the paper give a statistical picture of its preferred heading. This setup made it possible, in the 1960s and 1970s, to run careful cue-conflict experiments: change one cue while holding others steady, and see which way the bird insists is correct.

Those experiments revealed something that seems almost philosophical. Birds do not treat all cues as equal truths. They *rank* them, and the ranking shifts with circumstance. Under a clear sky, the stars can dominate nocturnal orientation. At sunset, patterns of polarized light (a subtle alignment in skylight, strongest near the horizon) can set the calibration of other compasses. Under heavy cloud, birds often rely more on magnetic direction. The sense of a hierarchy becomes obvious when cues disagree. If the magnetic field is experimentally rotated while the sky remains natural, some birds follow the manipulated magnetic direction; under other conditions they follow the sky and ignore magnetism. The point is not that birds are fickle, but that cue reliability is conditional. A compass that depends on seeing the horizon is a fine instrument until the horizon disappears.

The calibration story is where redundancy becomes more than mere backup. A backup compass is helpful; a compass that can *teach* another compass is transformative. Frank Moore and John Phillips, synthesizing many studies in the early 2000s, highlighted how often cue-conflict results look like calibration rather than simple dominance. One recurring outcome is that polarized light at sunrise or sunset can recalibrate a bird's magnetic compass: after exposure, the bird behaves as though magnetic north has been redefined to agree with the sky's geometry. Under other experimental geometries—for instance, when the horizon is blocked so that the bird sees only a patch of sky overhead—magnetic cues can appear to dominate, perhaps because the sky compass is missing its most informative region. Redundancy, in other words, is not just having multiple sensors; it is having rules for when to trust them and how to align them.

A human analogy helps. Suppose your phone's map says you are on the east side of a river, but you can see the river to your left. You do not average the two; you decide which is wrong. If you conclude the GPS is drifting, you use vision to correct it. But later, in thick fog, you may lean on GPS again because vision has become unreliable. Animals face the same logical problem, except their "GPS" might be magnetic, their "vision" might be stars, and their river might be an odour gradient.

Homing pigeons bring the map side of redundancy into sharper focus, and they do it with a plot twist that still unsettles intuition. For a long time, the dominant human story about pigeon homing was visual: pigeons recognize the landscape, remember routes, and make their way home the way a person might. That is certainly part of the truth near familiar areas. Yet decades of research have made it hard to escape a second ingredient, especially for unfamiliar release sites: smell.

The most famous strand of pigeon work is associated with Floriano Papi and collaborators in the 1970s and after, who argued that pigeons build an "olfactory map"—a learned association between wind-borne odours and directions, allowing a bird to infer where it is relative to home. The basic idea is surprisingly down-to-earth. Air masses carry mixtures of volatile chemicals that vary regionally: vegetation, soil, industry, sea spray, urban emissions. A pigeon at its loft experiences these odours arriving on winds from different directions. Over time it can learn a correspondence: *this smell tends to come from the north, that smell from the southwest.* Later, released elsewhere, the bird can compare local odours to its learned template and infer which direction home lies. If that sounds too humanly cognitive, consider what the alternative would be: a single global coordinate system, precise enough to locate a loft, delivered by an invisible field alone. The olfactory story may be messy, but it matches the messiness of real air.

The strongest evidence for an olfactory contribution is not poetic; it is surgical and logistical. In some experiments, pigeons deprived of normal olfactory input (by treatments that block the sense of smell) show impaired homing from unfamiliar sites, while still showing normal motivation and flight ability. Other manipulations, such as transporting birds in air filtered to remove environmental odours, can disrupt their ability to determine the correct initial homeward direction. None of this implies that smell is the only map. It implies that pigeons treat odour as a serious source of position information when landmarks are unfamiliar.

Now add compasses to that map. A pigeon can have a good sense that home lies, say, west-northwest. To *fly* west-northwest, it needs a heading reference. Under a bright sun, the sun compass can serve. Under overcast, a magnetic compass becomes useful. Near home, familiar landmarks can dominate and compress uncertainty to nearly zero: a church spire, a highway, a coastline. The redundancy is layered: map-like odour information to select a bearing, compass information to hold it, and visual piloting to finish. The same bird can succeed even if one layer is weakened, which is exactly what makes the system hard to understand by tinkering with only one cue at a time.

Redundancy also shows up in an uncomfortable way: animals can look irrational when the experimenter removes a cue the animal would not normally lose. In a laboratory cage, a bird cannot use landmarks. In a funnel, it cannot fly forward and sample changing odour fields. A cue that is minor in the wild may become dominant in the lab simply because everything else has been stripped away. This is not a flaw; it is a warning label. "Dominance" in cue-conflict tests can be a property of experimental geometry as much as the animal's priorities. Block the horizon and you cripple parts of the sky compass. Add metal near the cage and you distort magnetism. The animal's brain is then forced to do what brains always do: make the best of the information it has.

There is a second reason redundancy matters, and it lies deep in the physics of cues. Many natural gradients are noisy. Odours disperse turbulently. Winds shift. Magnetic intensity changes smoothly over hundreds of kilometres but can be perturbed locally. Celestial cues depend on time and weather. A reliable navigator needs not just multiple cues, but multiple cues that fail *differently*. When cloud blocks stars, odour still exists. When a sudden shift in wind scrambles odour, magnetism remains. When magnetism is locally distorted near a basalt outcrop or a steel bridge, the sun is indifferent. Redundancy works because the world's errors are not perfectly correlated.

This is also where debates become informative rather than irritating. In birds, the existence of a magnetic *compass* is well supported by decades of work. The existence of a robust, bicoordinate magnetic *map* in birds is more contentious. Critics point to the complexity of the crustal magnetic field and the weakness of large-scale gradients at the spatial scale a bird would need for pinpoint navigation. Others argue that birds could use magnetism more modestly: as a signpost near coastlines, as a way to detect latitude-like changes, or as one ingredient among others, perhaps combined with learned geography. Some researchers have proposed alternative or auxiliary map cues such as infrasound—low-frequency sounds generated by ocean waves and terrain interactions that can propagate long distances. Whether or not infrasound plays a central role, the broader lesson is that a map does not have to be monogamous. The very fact that there are plausible candidates beyond magnetism hints at how natural selection hedges its bets.

A historical anecdote makes that hedging feel less abstract. During the era when pigeon post was a serious military tool, lofts were protected like infrastructure. The whole point of the system was reliability under adverse conditions: smoke, haze, disrupted roads, damaged communications. Trainers did not assume a single cue would always be available. They released birds in varied weather and from varied directions. That practice, driven by wartime necessity rather than academic curiosity, anticipated what laboratory science later formalized: navigation is a suite of tricks, not a single trick.

A subtle form of redundancy is *learning what not to trust.* Animals exposed repeatedly to a cue that proves misleading can reduce its weight. This is hard to demonstrate cleanly, but the logic is unavoidable. A young migrant experiences the sky, the magnetic field, wind drift, and the consequences of its own steering. Over time, correlations between cues can be used to calibrate them. The magnetic compass, for example, offers a stable reference that is available even when the sky is hidden. But the magnetic field is not labelled "north" in any absolute sense; it is a directional axis that must be interpreted. If polarized light at sunset provides a geometrically meaningful reference, then aligning the magnetic axis to that reference makes sense. Developmental studies that manipulate magnetic conditions during early life and then test orientation later suggest that magnetic information can influence how a bird's compass system is set up, as if the animal is building a consistent internal coordinate frame from multiple partial truths.

That idea can be brought back down to a scene anyone can picture: a young bird on its first migration. The sunset is visible, and the pattern of the sky has a dependable geometry. The magnetic field is also present, invisible but steady. If the bird can align those two references early on, then later, on an overcast night, magnetism can stand in for the missing stars. Redundancy has become inheritance plus experience: an innate sensitivity tuned by the world.

All of this complicates the temptation to ask, "Which sense is the real one?" The better question is, "Which sense is reliable *now*, and how does the animal know?" Sometimes the answer may be as simple as an evolved priority: under thick cloud, follow magnetism. Sometimes it may be as subtle as a context cue: near home, trust landmarks; far away, trust the map. Sometimes it may be a memory of recent agreement: if the sun and magnetism have matched consistently for days, then an abrupt mismatch suggests one is being distorted.

The magnetic sense, in this redundant picture, is neither an all-powerful explanation nor a minor accessory. It is an unusually democratic cue: available day and night, over land and sea, in fair weather and foul. That makes it an excellent compass backup and a plausible contributor to map-like information in some environments. But its very availability can also create traps. A magnetic cue that is normally reliable can become misleading in a local anomaly, and an animal that leans too heavily on it can be steered wrong. The best navigators are not those with the strongest single sense; they are those that can *notice* when a sense has become suspicious.

A final image, simple enough to hold in the mind, captures what redundancy buys an animal. Think of a bird leaving a release site it has never seen before. The air carries a certain chemical mix; the bird compares it to a learned association and infers a rough homeward direction. It sets that direction using the sun if the sun is visible, or the magnetic field if it is not. As it flies, the odour blend shifts; the map estimate is updated. The clouds thin; stars appear; the bird checks that its magnetic heading is consistent with the sky and quietly tweaks the calibration. Nearer home, a familiar river appears, and the last uncertainty collapses into confident flight along a remembered corridor. No single step is miraculous. The miracle is that failure in one channel does not force failure of the whole, because the animal's world is read through several overlapping senses, each correcting the others just enough to keep home within reach.

## 2.3 The Helmholtz Deception

The Earth's magnetic field is a strange experimental enemy. It is everywhere, it passes through bodies and buildings, and you cannot put it in a box and keep it out the way you can exclude light or sound. If an animal turns north in a laboratory, a sceptic can always say it was following a draft, a faint glow, a memory, the smell of the researcher's lunch. The magnetic cue is invisible, so the burden of proof is heavier: to convince anyone, you have to show not merely that an animal *can* orient, but that it *changes* its behaviour when the magnetic field is changed, while everything else stays the same.

That is the logic behind the most elegant trick in magnetoreception research: don't try to block the field; *override* it. Replace nature's magnetic instructions with your own. If the animal follows the replacement, the magnetic sense becomes hard to deny.

The name most often attached to this deception is Hermann von Helmholtz, though the device itself belongs to an engineering tradition that long outgrew any one person. A pair of circular coils, arranged parallel to each other and separated by a specific distance, can generate a region of magnetic field that is remarkably uniform. Drive a controlled electric current through the coils, and you can create a field pointing in any chosen direction. You can add that field to the Earth's field, partially cancel the Earth's field, or rotate the combined field as if you had taken the entire planet's magnetism and turned it like a steering wheel. In the jargon, these are Helmholtz coils.

The coils solve a practical problem and a philosophical one. Practically, they create an experimental arena in which the magnetic environment is measurable and adjustable. Philosophically, they let you perform a causal intervention on an otherwise untouchable variable. When scientists debate whether an animal really uses magnetic information, they are debating causation. Correlation is cheap: a bird scratches toward north and a compass on the wall also points north. Causation requires the cruelty of a good magic trick: secretly rotate north, and see whether the bird rotates with it.

This approach is most famous in bird orientation experiments, partly because the behaviour is so easy to record and partly because the animals are so motivated. A nocturnal migrant in autumn, restless in a cage, is a natural compass needle. Place such a bird in an Emlen funnel, and you can measure its preferred direction. Now put the entire funnel inside a set of coils and rotate magnetic north by, say, 90°, while keeping the room dark and featureless. If the bird's scratches rotate by 90° as well, it is difficult to argue that magnetism is irrelevant. The bird has been tricked, and the trick has a clean signature.

The beauty of the method is that it can be scaled from gentle to extreme. You can rotate the horizontal component of the field while keeping strength constant. You can change inclination (the dip of the field lines) while leaving the compass direction intact. You can reduce the field to near zero, creating a magnetic "silence" that many animals find disorienting. You can also create deliberate conflicts: keep the sky cues natural but rotate the magnetic field, or keep magnetism natural but rotate a pattern of polarized light in an artificial sky. These conflicts are not mere cruelty; they are questions. Which cue wins? Under what conditions? Does the animal treat one cue as a reference that calibrates the other?

At this point the coils become more than a lie detector; they become a way to map a mind. If an animal follows magnetism when the sky is unavailable, but follows the sky when it is clear, that suggests a hierarchy of cue use. If an animal initially follows the sky and later behaves as if its magnetic compass has been reset to agree with it, that suggests calibration. Frank Moore and John Phillips emphasized how such cue-conflict experiments can show asymmetric relationships: one cue may *set* the other rather than simply overpower it. A coil system is, in that sense, a tool for studying not just magnetism, but decision-making under uncertainty.

The method has a pleasingly old-fashioned experimental flavour: if you can't see the variable, build an apparatus that forces it to reveal itself. There is a historical echo here of early electricity demonstrations in which invisible charges became visible through sparks and deflected needles. In magnetoreception, the deflected needle is the animal.

What makes the deception particularly convincing is that it can be paired with controls that anticipate the sceptic's objections. Coils can be run with currents that produce heat, vibration, or faint noise, so a careful design includes sham conditions in which current flows but the net magnetic field change is cancelled. Researchers can rotate the coils physically without changing the field, to test whether the animal responds to the apparatus rather than the magnetism. The goal is to isolate the magnetic variable so completely that the only plausible explanation for a behavioural shift is a change in magnetic information.

The logic is simple enough to explain over dinner, but the execution is not. Creating a uniform field requires careful geometry. Measuring it requires magnetometers and calibration. Shielding the room from stray electromagnetic noise can matter, because some proposed mechanisms for magnetic sensing are sensitive to oscillating fields in the radiofrequency range. Even the wiring matters. A poorly designed setup can create gradients that allow animals to use intensity differences as a cue when you meant to test direction, or to use unintended artefacts like heating. The coils are not a magic wand; they are a disciplined way of lying.

One concrete example of deception that feels almost like science fiction comes from the way coils can create *virtual locations* for animals that never move. In marine navigation studies, turtles or salmon can be placed in tanks or flumes and exposed to magnetic fields that match those found hundreds or thousands of kilometres away. The room does not change. The water does not change. The animal's only "travel" is through magnetic space. If the animal responds as though it is in a different geographic region—choosing a direction appropriate to the simulated location—that is strong evidence for map-like use of magnetic parameters. The point is not merely that the animal has a magnetic compass, but that it can interpret a particular combination of inclination and intensity as an indicator of position. Coil systems make that test possible because they hold everything else constant.

There is a deeper conceptual advantage here. Displacement experiments with real animals in the field are always messy: transport may introduce odours, sounds, vibrations, and stress. The animal might glimpse landmarks. Weather changes. In a laboratory magnetic displacement, the animal remains physically where it is, eliminating a raft of confounds. You are displacing the *cue*, not the creature. If the animal behaves differently, it is responding to the cue itself.

The same apparatus can also expose the limits of magnetic navigation. If an animal's response vanishes when you cancel the field, it suggests magnetism is necessary under those conditions. If the response persists, it suggests magnetism is optional, or that another cue has taken over. Either outcome is informative. A negative result, in a well-controlled coil system, can be just as important as a positive one: it tells you where to stop attributing powers to magnetism.

The deception becomes even more interesting when it is extended across time, not just across a single test. A one-off rotation of magnetic north tells you whether the animal can sense and use the field in the moment. But what happens if you change the magnetic environment during development, when the animal's navigational reference frames are being built? Helmholtz coils have been installed at nest boxes so that young birds experience a shifted magnetic field during incubation or nestling stages. Later, when those birds are tested during migratory restlessness, their preferred orientation can be shifted in a way consistent with the early manipulation. The implication is provocative: magnetism may serve as a foundational reference that helps the animal learn how other compasses map onto the world. In such a view, the magnetic field is not merely another cue in the adult's toolkit; it is part of the scaffolding on which the toolkit is assembled.

That developmental angle also guards against a seductive misunderstanding. When people hear that birds use stars or the sun, they may imagine a celestial compass as the primary, noble guide, with magnetism as a grubby backup. But a young bird cannot use stars effectively without learning which stellar pattern corresponds to which direction, and that learning requires some stable reference. Magnetism is available even before a bird ever leaves the nest. It makes sense, therefore, that magnetism could act as a training wheel for more visually rich systems. The coils let you test this, because you can alter the training wheel without altering the sky.

At the level of mechanism, the coil deception can also discriminate between different hypotheses about how magnetism is sensed. Two broad families of ideas dominate discussion. One is that magnetism is detected through a light-dependent chemical process, often associated with the protein cryptochrome in the retina. In this view, magnetic information is linked to vision-like processing and can be disrupted by particular radiofrequency fields that perturb delicate spin dynamics. The other is that magnetism is detected through tiny magnetic particles (commonly discussed as magnetite) coupled to the nervous system, potentially providing information about intensity or direction through mechanical forces. These are not mutually exclusive; different mechanisms could serve different functions, such as compass versus map. But because the field itself cannot be easily eliminated, teasing mechanisms apart requires carefully designed perturbations, and coils provide the baseline control necessary for such perturbations to mean anything.

A striking recent example comes from conditioning work with sea turtles that turns magnetic map research into something closer to a learning assay. Rather than merely observing spontaneous orientation, researchers can associate particular magnetic signatures with a reward, such as food, and test whether turtles learn to distinguish those signatures. Once you have a learned response, you can probe it with additional manipulations: change magnetic parameters slightly; add noise; introduce oscillating radiofrequency fields expected to disrupt radical-pair chemistry. Reports that such radiofrequency treatments did not disrupt the conditioned magnetic-map responses in turtles place constraints on candidate mechanisms for that map sense. The experimental philosophy is the same as in classic coil deception: if you can control the invisible variable precisely, you can ask sharper questions about what the animal is actually using.

There is also a practical reason this deception has become so central. Many animals that are magnetic-sensitive are small, fragile, or ethically challenging to test in the wild in large numbers. A coil system allows repeated tests under safe, controlled conditions, reducing stress and improving statistical power. It also enables experiments that would be impossible outdoors, such as presenting an animal with a magnetic field corresponding to a location it has never visited, or presenting fields in rapid sequence to test discrimination thresholds.

Even so, the coils do not abolish ambiguity; they move it. An animal might respond to the horizontal direction of the field, or to inclination, or to intensity gradients within the coil system. It might respond differently depending on light conditions, suggesting a light-dependent compass. It might show responses that change with season, age, or motivational state. The deception tells you that magnetism can influence behaviour, but it does not, by itself, tell you the ecological context in which magnetism matters most. That requires field work, tracking, and an appreciation of how animals combine cues in the messy real world.

The most satisfying coil experiments are those that keep the magic trick honest. They specify the behavioural prediction in advance: rotate north by 90°, and a compass-based heading should rotate by 90°. Cancel the field, and orientation should weaken or scatter if magnetism is required. Present a field characteristic of a different region, and a map-like response should shift accordingly. When those predictions are met, the invisible becomes testable.

There is something almost endearing about the way this line of research forces humility. Humans have a hard time accepting senses we do not share. We trust what we can see and hear. The coil apparatus is a reminder that nature is under no obligation to fit inside our sensory envelope. An animal can be loyal to a cue we cannot feel, and the only way to respect that fact is to give ourselves the power, briefly and carefully, to change the cue and watch the animal notice.

A set of coils humming quietly in a darkened room does not look like the stuff of grand biological revelations. Yet in that controlled bubble of altered magnetism, a bird's restless hops or a turtle's steady swimming can become an argument that is hard to refute: the world has an invisible geometry, and some nervous systems have learned to read it.

Chapter 3

# A Quantum Compass in the Eye?

*Deep within the retina of a migrating robin, a fleeting chemical reaction harnesses the strange laws of quantum mechanics to detect the Earth's weak magnetic field. We explore how light-sensitive proteins called cryptochromes may act as a biological heads-up display, translating invisible magnetic lines into tangible chemical signals.*

## 3.1 The Light-Dependent Switch

A compass that only works when the lights are on sounds like a gimmick. Yet for many migratory birds it is a stubborn, repeatable fact: change the colour of the light, dim it too far, or deprive the animal of short wavelengths, and the familiar directional preference can falter or vanish. That observation does more than narrow down the list of candidate sensors. It points, with unusual force, to a particular kind of biological hardware: a molecule that *needs* a photon to begin its work.

One of the most influential early clues came from a series of elegant behavioural experiments beginning in the 1960s, associated especially with Wolfgang and Roswitha Wiltschko. Their birds could orient using the Earth's magnetic field, but the "magnetic compass" was not like a tiny bar magnet. It behaved like something that depended on the surrounding conditions in ways a magnet does not. Over time, different laboratories refined the picture: it was not merely that birds preferred to travel at night or that their eyes were generally important for navigation. The orientation itself could depend on the spectrum of light available, with blue and green light often supporting magnetic orientation where longer wavelengths did not. That is the kind of constraint that makes biologists sit up: whatever the magnetic sensor is, it is tied to photochemistry.

Photochemistry is chemistry *started* by light. In everyday life it is the reason a photograph forms on film, why sunlight can fade fabric, and why plants can use energy from photons to build sugars. In a cell, photochemistry is usually carried out by proteins that hold onto a light-absorbing cofactor. The cofactor is the "business end" of the device: a molecule with electrons arranged so that one of them can be pushed into a more energetic configuration by the right photon. The protein is the scaffolding: it positions the cofactor, tunes its sensitivity, channels the energy into a controlled reaction, and links that reaction to the rest of the cell's signalling machinery.

The leading candidates for the light-dependent magnetic compass belong to a family of proteins called cryptochromes. Their name has a mildly theatrical origin. They were first recognised not as magnets but as "cryptic" photoreceptors: molecules that clearly responded to light in plants and animals, yet did not fit neatly into the better-known photoreceptor families. Today cryptochromes are familiar to chronobiologists because they play key roles in circadian timing in many organisms, including insects and mammals. But cryptochromes are also photochemists at heart, and that matters even when their biological job is not to keep time.

The simplest way to picture a cryptochrome is as a carefully shaped pocket holding a special cofactor: flavin adenine dinucleotide, usually abbreviated FAD. Flavins are the yellowish molecules behind the colour of riboflavin (vitamin $B_2$), and they are among biology's most versatile electron-handlers. In many enzymes, flavins act like rechargeable batteries, cycling through different redox states as they accept and donate electrons. Cryptochromes borrow this talent, then combine it with a crucial extra trick: the flavin can be excited by short-wavelength light, especially blue.

A single photon is not "a lot of energy" on human scales, but in molecular terms it is decisive. Absorb one blue photon and you can shove an electron into an excited state, creating a molecule that is eager to rearrange its charges. The protein, having evolved to exploit that brief moment, offers an electron donor nearby. In many cryptochromes the donors are tryptophans, bulky amino acids whose ring structures are good at letting electrons hop. Often the electron does not move in one jump but along a short stepping-stone chain, sometimes described as a "tryptophan triad" or "tetrad" depending on the species and the details of the pathway.

The consequence of that electron transfer is easy to state and surprisingly hard to digest. The flavin becomes a radical, and the donor becomes a radical too. A radical is a molecule with an unpaired electron. We are used to radicals as villains in chemistry: reactive, unstable, the stuff of oxidative damage. In cells they can indeed be dangerous. But in controlled settings radicals can also be exquisitely informative. They are like molecular needles that can sense subtle influences, because an unpaired electron carries a magnetic moment and has a quantum property called spin.

When the photon-triggered electron transfer produces two radicals at once, it creates what chemists call a radical pair. The phrase sounds like a mere bookkeeping convenience, but it marks a special state of affairs: two unpaired electrons, created together in a linked quantum state, embedded in a protein that constrains how they move and what reactions they can undergo. This radical pair is the heart of the proposed magnetic sensor. The critical point for now is simpler than the quantum details: without light, no excited flavin; without excited flavin, no electron transfer; without electron transfer, no radical pair. The compass, in that sense, has a switch, and that switch is a photon.

That "switch" must also be reset. A sensor that turns on once and stays on would be a poor compass. Cryptochromes therefore operate in cycles, moving between a resting state and one or more active states, then returning again. In broad terms, the resting cryptochrome holds FAD in an oxidised form. Blue light excites the flavin, an electron is transferred in, and the flavin becomes reduced while the donor becomes oxidised: two radicals appear. After a short lifetime the radicals recombine or proceed along distinct chemical pathways, producing downstream changes in the protein's shape or its interactions with partners, and eventually the system re-oxidises, ready for another photon.

This cycle is not merely a chemical curiosity. It addresses a biological necessity: a bird in flight does not get a single reading from its compass. It samples continuously, with the output integrated over time by neural circuits that must remain stable across minutes and hours. A photochemical cycle provides a natural rhythm of sampling. Each absorbed photon is like a tiny "measurement event," briefly creating a sensitive state that can be nudged one way or another, then collapsing back into a resting configuration. Multiply that by the enormous number of cryptochrome molecules in a retina and the result could be a steady stream of information.

It is worth pausing over the setting: the retina. The eye is not just a camera. It is a layered sheet of neural tissue that happens to be light-sensitive. In addition to rods and cones, which perform the familiar job of vision, the retina contains many other proteins tuned to light for different purposes, from regulating pupil size to synchronising the body clock. If a magnetic compass depends on light, the retina becomes an obvious place to look, not because it is mystical, but because it is already the body's most sophisticated laboratory for photon-driven chemistry.

A concrete example of how this molecular idea meets biology comes from the European robin, a small songbird whose migratory restlessness has made it a favourite in orientation studies. The robin's magnetic compass can be studied in controlled conditions: place the bird in a special cage that records its scratching direction, adjust the magnetic field with coils, vary the light spectrum, and watch whether the animal's preferred direction tracks the manipulated magnetic cues. This behavioural work helped motivate the search for specific cryptochrome variants in bird retinas that might plausibly act as magnetically sensitive photoreceptors. The attention has increasingly focused on a protein called cryptochrome 4 (Cry4), which appears in the retina and, in several studies, shows patterns consistent with a role in migration-related orientation rather than only daily timekeeping.

At the molecular level, Cry4 behaves like a flavoprotein should: it binds FAD and can undergo light-driven electron transfer. In 2021, a striking biochemical result strengthened the plausibility of the entire picture: purified cryptochrome 4 from a migratory songbird showed reaction yields that changed with applied magnetic fields, and mutations along the electron-transfer pathway altered that sensitivity. The significance is not that a test tube can "navigate," of course. The significance is that a single protein, isolated from the animal, can display magnetic-field-dependent chemistry under conditions relevant to the Earth's field. That is precisely what a light-dependent magnetic compass requires at its base: a molecule that can translate photons into a spin-sensitive radical pair, and radical-pair dynamics into different chemical outcomes.

Even so, cryptochrome should not be imagined as a tiny compass needle. It is better pictured as a molecular decision point. Light creates a reactive intermediate, and the subsequent fate of that intermediate can take more than one route. The routes differ in the chemical products they yield, in the duration of intermediate states, and in how long the protein stays in an "activated" configuration. In many signalling proteins, these differences are what downstream partners detect. A receptor might bind a partner only in one redox state; an enzyme might expose a docking site only when a cofactor is reduced; a channel might open only when a domain shifts position. In cryptochromes, the details vary across species, but the general theme holds: the photochemistry is translated into changes that other proteins can read.

One can understand the appeal of this mechanism by thinking about what a nervous system needs. A compass must provide a directional cue that is stable enough to guide behaviour yet flexible enough to integrate with other senses. The retina offers a computational advantage: it can represent patterns across space. If cryptochrome molecules are arranged or anchored in a consistent orientation within retinal cells, then the outcome of their photochemistry could vary systematically across the retina depending on the angle between the molecule's internal axes and the geomagnetic field. The bird would not "see north" as a bright arrow, but the visual system could, in principle, detect a faint, field-dependent modulation overlaying the normal visual scene: a pattern whose symmetry reflects the inclination of the field lines rather than their polarity.

That suggestion makes a subtle prediction about light. Cryptochromes are tuned to shorter wavelengths not because birds prefer blue aesthetically, but because the photon must have enough energy to drive the flavin into its excited state and initiate electron transfer efficiently. Blue and near-UV light are well-suited to flavin absorption. Red light, with its longer wavelength and lower photon energy, is often less effective at producing the key excited state. In practice, the boundary is not a sharp line, because proteins, cofactors, and cellular conditions can shift absorption spectra. But the broad match between flavin photochemistry and the behavioural dependence on short-wavelength light has been one of the most persuasive pieces of circumstantial evidence for cryptochrome involvement.

The same light dependence also helps distinguish cryptochrome-based ideas from another major candidate mechanism: magnetite. Magnetite is an iron oxide that can form tiny crystals with permanent magnetic properties. If an animal had magnetite particles mechanically linked to stretch-sensitive channels, the Earth's field could tug on them and produce a sensory signal without requiring light. Many animals almost certainly do use magnetite for some magnetic tasks, especially those involving intensity rather than direction. But the fact that the avian compass can be light-dependent, and in some experiments seems tuned to particular wavelengths, makes a purely magnetite compass an awkward fit for those behaviours. Cryptochrome, by contrast, *expects* light dependence as a fundamental feature.

Still, the phrase "light-dependent switch" should not tempt us into thinking the biology is a simple on/off button. Real retinas are messy optical environments. Light intensity varies by orders of magnitude between dusk and moonlit night. Birds migrating at night see dim, blue-shifted skylight; birds active by day encounter intense, broad-spectrum illumination. A photochemical receptor must avoid saturating in bright light, must reset in darkness, and must remain informative across different photic conditions. This is one reason cryptochrome cycles have attracted such careful attention. Some behavioural findings suggest that the crucial magnetically sensitive step may occur during a re-oxidation phase that can proceed in darkness after initial light exposure. That idea, if correct, would reconcile a light requirement (the system needs to be primed) with the possibility that the most sensitive radical pair state might arise not during the instant of photon absorption but during subsequent chemical housekeeping.

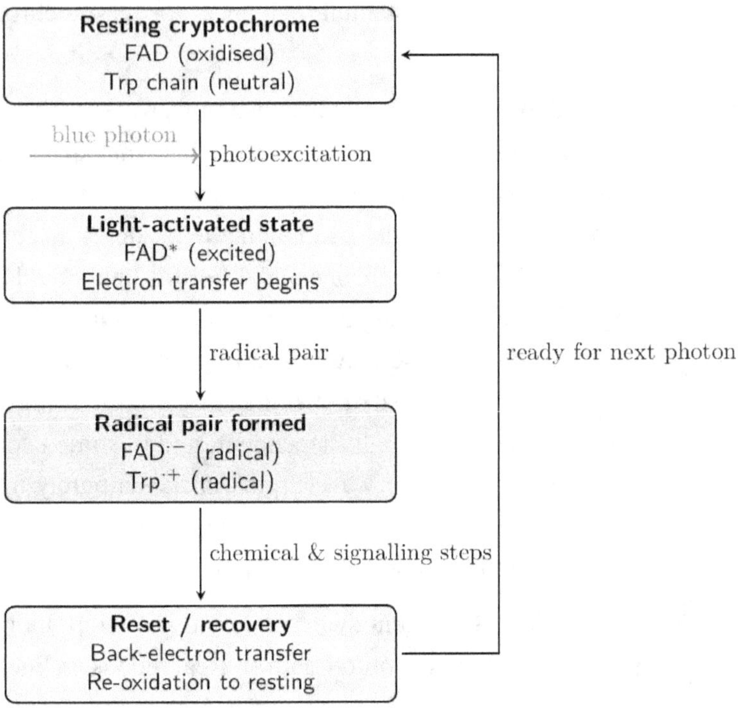


There is another reason cryptochromes have become such compelling suspects: evolution has already "solved" the problem of putting photochemistry into sensory pathways. Vision itself is built on a photochemical switch. A retinal molecule inside opsin absorbs a photon, changes shape, and triggers a cascade that ultimately alters an ion channel. The key difference is that opsins rely on a large conformational change in a bound chromophore, while cryptochromes rely on electron transfer chemistry around a flavin. But the logic is the same: a tiny quantum event (absorption of one photon) can be amplified into a macroscopic behavioural response.

If cryptochrome is involved in magnetic sensing, one of the hardest biological questions is also one of the most basic: how does a cryptochrome molecule become a compass *sensor* rather than simply a light sensor? The answer must involve geometry. A compass needs direction, and direction requires some kind of orientation reference inside the body. A freely tumbling molecule in a cell would average away directional information. The protein therefore has to be constrained: anchored in a membrane, arranged in an ordered array, or held in a cellular structure that fixes its axes relative to the retina. This is not an abstract concern; it is a design requirement. A molecule can be exquisitely sensitive to magnetic effects and still be useless for a compass if its orientation is random.

Biology often meets such requirements through sheer numbers and modest biases rather than perfect alignment. The retina is packed with elongated cells and layered membranes. Many proteins are targeted to specific compartments. Even a partial ordering could be enough if the nervous system can integrate the resulting pattern across many receptors. The needed degree of order remains a matter of active debate, and it is one reason the story has not yet reached the satisfying simplicity of "here is the exact receptor, here is the exact circuit, here is the exact percept." But the photochemical foundation is unusually solid: cryptochromes are real, flavins are real, blue-light electron transfers are real, and radical pairs are a natural consequence of that chemistry.

For a migrating bird, the practical outcome is that the sky is not merely a backdrop of stars. It is also the energy source that primes a molecular compass. Under the right light, a protein in the retina becomes briefly, repeatedly, delicately charged with two radicals whose fates can be biased ever so slightly by the geomagnetic field. The bird does not need a magnet in its head; it needs a photochemical switch in its eye that can be flipped millions of times, reliably, as it moves through night air and changing weather, turning light into a signal that the brain can learn to trust.

## 3.2 The Quantum Tug-of-War

A radical pair is born in a very particular mood. Two unpaired electrons appear at once, created by a single electron transfer, and for a fleeting moment their properties are correlated in a way that ordinary, slowly heated chemistry almost never produces. The correlation is not a vague "they influence one another" but a specific quantum relationship between their spins.

Spin is an awkward word, because it invites the wrong picture: two tiny balls twirling like tops. Nothing is actually spinning in space. Spin is a built-in quantum property of electrons, more like a label that determines how they behave in magnetic fields and how two electrons can be arranged together. Each unpaired electron acts, in effect, like a minuscule bar magnet with a north and south pole. Put it in a magnetic field and its preferred energy depends on how that "magnet" is oriented. That much is already enough to make you suspicious that electrons could be magnetic sensors. The difficulty is scale. The Earth's magnetic field is weak. Thermal motion in warm, wet tissue is relentless. How could such a small field have any reliable effect?

The counterintuitive answer is that the field does not have to overpower thermal noise. It only needs to *steer* a competition that is already delicately balanced by the rules of quantum mechanics. Radical pairs provide exactly that kind of competition: a tug-of-war between two possible spin arrangements, each one connected to different chemical fates.

The two arrangements have names that sound like characters in an opera: singlet and triplet. The names come from spectroscopy, but the idea can be explained without equations. Take two electrons. Each electron has spin that can be thought of, loosely, as "up" or "down" along any chosen axis. When two unpaired electrons are considered together, their spins can combine in two qualitatively different ways. In a *singlet* state, the two spins are paired oppositely in a way that makes the combined magnetic moment cancel. In a *triplet* state, the spins are aligned in ways that produce a net magnetic character. The crucial detail is not just that the labels differ, but that singlet and triplet radical pairs tend to follow different chemical pathways, because the rules for how electrons can pair up again (and therefore which bonds can form or break) depend on the spin state.

Imagine the radical pair as two dancers on a crowded floor, briefly linked by the fact that they began the dance together. If the pair stays in a singlet state, the electrons are "compatible" for recombination: they can come back together into a paired-electron bond and return the molecule to a quiet, closed-shell condition. If the pair is in a triplet state, recombination may be hindered or redirected. The radicals might persist longer, react with other molecules, or lead the protein into a different conformation. The chemistry downstream becomes a branching pathway, and the branch taken depends on how long the radical pair spends in singlet versus triplet character.

Nothing about this yet requires the Earth. The Earth enters because singlet and triplet are not always fixed identities. In many radical pairs they interconvert, oscillating between singlet-like and triplet-like character. This oscillation is driven by internal magnetic interactions within the molecule: the tiny magnetic fields produced by nearby atomic nuclei (especially hydrogen and nitrogen) and by the electrons themselves. Those internal interactions vary with molecular structure, which is why the details of a cryptochrome's flavin and its electron donor chain matter so much. They set the rhythm of the oscillation.

Now add an external magnetic field, even a weak one. The field slightly shifts the energy levels associated with different spin orientations, and it changes the way the two-electron system precesses, the way a tilted spinning top wobbles in a gravitational field. Again, "precession" is a metaphor, but it captures the key point: the quantum state evolves in time, and the magnetic field changes that evolution. The effect can be tiny at any single instant, yet meaningful when integrated over the lifetime of the radical pair and over countless repetitions. A small bias in the fraction of time spent as singlet rather than triplet can translate into a measurable change in reaction yields, and therefore into a biological signal.

It helps to fix the timescales, because otherwise "tiny" is too vague to take seriously. Under an Earth-strength magnetic field, an electron's spin precesses at a frequency on the order of a megahertz, meaning a period of roughly a microsecond. That is astonishingly fast compared with human intuition, but not absurdly fast for chemistry. Many radical pairs in proteins can live for microseconds or longer, especially if the protein structure keeps the radicals spatially separated enough to slow their recombination. If a radical pair remains coherent, maintaining its quantum spin correlation for at least a comparable timescale, then the Earth's field has time to bend the evolution one way rather than another. That is the narrow window where a planet's field can become a chemical influence.

The word *coherence* is where many people's patience snaps. It sounds like a fragile, laboratory-only concept, the sort of thing that requires superconducting magnets and cryogenic temperatures. In the radical-pair mechanism coherence is not an optional philosophical flourish. It is simply the statement that the radical pair is created in a well-defined quantum state and evolves under quantum rules for long enough that singlet and triplet character can mix in a meaningful, field-dependent way. Warmth does tend to destroy coherence, but the destruction is not instantaneous. A protein can shelter its radicals, and more importantly, the radical pair does not need to remain coherent for seconds. Microseconds can be enough. In a sense, the protein is not trying to build a quantum computer; it is trying to keep a quantum coin spinning just long enough for the Earth to nudge the odds before the coin lands.

A historical detour makes this less alien. In the 1970s and 1980s, chemists studying reactions in solution noticed puzzling magnetic-field effects. Some photochemical reactions produced different product ratios when performed in magnetic fields, even modest fields, and the changes could not be explained by classical forces on charged particles. This was the birth of spin chemistry as a field. Researchers learned that radical pairs created by light often began in a singlet state and that their subsequent interconversion with triplet states could be influenced by magnetic fields. The radical-pair mechanism for animal magnetoreception did not fall from the sky fully formed; it was an audacious borrowing from a body of chemical physics that already knew radical pairs were magnetically sensitive. What was new, and daring, was the claim that evolution might have built a biological receptor that harnessed those effects deliberately.

The question then becomes: how does a radical pair convert magnetic direction into something directional, rather than merely "field present" or "field strong"? The trick is anisotropy: direction-dependent behaviour that arises because the molecule is not spherical. Inside a cryptochrome, the radicals occupy fixed positions and experience internal hyperfine interactions with nearby nuclei in a way that depends on the orientation of the molecule relative to the magnetic field. Rotate the molecule, and the pattern of spin mixing changes. If many molecules are similarly oriented in the retina, the reaction yield becomes a function of magnetic direction, not just magnetic strength. The nervous system could, in principle, read out that directional dependence as a pattern.

This is where earlier behavioural observations acquire an unexpectedly precise meaning. Birds' magnetic compass often acts as an inclination compass: it is sensitive to the *tilt* of the field lines (the angle they make relative to the Earth's surface), not to the polarity that distinguishes magnetic north from south. Turn the field upside down and the animal's preferred direction tends to reverse in a way consistent with "poleward" versus "equatorward" rather than "north" versus "south." That behaviour is awkward for a compass needle made of magnetite, which naturally distinguishes polarity. But it drops out naturally from a radical-pair sensor, because the physics depends on the axis of the field, not its sign. In the simplest versions of the radical-pair mechanism, reversing the field is equivalent to a rotation by 180°, and many reaction yields are insensitive to that reversal. The chemical compass is a compass of axes, not arrows.

The tug-of-war metaphor earns its keep here. Imagine singlet and triplet pathways as two teams pulling on the same rope, each trying to claim the fate of the radical pair. The Earth's field does not grab the rope and yank. It changes the footing, the friction, the tiny timing differences that determine which side gains a slight advantage over thousands of rounds. The protein, meanwhile, acts as both arena and referee, setting the distances between radicals, the placement of nuclei that generate hyperfine interactions, and the availability of reaction partners. Under one orientation the singlet pathway may win a little more often; under another orientation the triplet pathway may. If the downstream signalling depends on which product is made, the molecule becomes a directional sensor.

A concrete example of how sensitive this can be comes from work on cryptochrome 4 in migratory songbirds. Purified protein, illuminated under controlled conditions, shows reaction kinetics and yields that can be altered by applied magnetic fields. Even more telling, mutating particular amino acids along the electron-transfer chain changes the magnitude of the effect. That is an experimental fingerprint of the underlying mechanism. It says: the radical pair is not an incidental side-effect; it is embedded in a particular electron pathway, and tinkering with that pathway alters magnetic sensitivity. The protein behaves like a device whose sensitivity is designed into its wiring.

At the same time, the experiment also brings the unresolved problem into focus. A change in product yield is not, by itself, a signal a neuron can read. A neuron reads changes in membrane voltage and neurotransmitter release. Somewhere between the chemistry in cryptochrome and the electrical language of the retina there must be a transduction step. Perhaps the cryptochrome, in one redox state, binds to a partner protein that initiates a cascade. Perhaps it modulates ion channels indirectly. Perhaps it influences the metabolism of reactive oxygen species in a tiny compartment, with downstream effects on signalling pathways. Several ideas circulate, and some proposed interaction partners have been identified in exploratory screens, but the chain from radical pair to retinal output is still incomplete.

That incompleteness should not be confused with flimsiness. Biology often reveals itself in layers. First comes a behavioural phenomenon: a compass that depends on light and responds to subtle manipulations of the magnetic field. Then comes a plausible molecular candidate: a flavoprotein in the retina capable of forming radical pairs. Then comes a physical mechanism that can, at least in principle, connect Earth-strength fields to chemical outcomes: singlet—triplet interconversion. Only later do the full cellular pathways come into view, and they can take decades.

One reason the radical-pair idea has retained its grip is that it makes distinctive, almost theatrical predictions. A chemical compass based on spin dynamics should be vulnerable to oscillating magnetic fields in the radio-frequency range, because such fields can drive transitions between spin states or scramble the coherent evolution. Classical magnetic particles, by contrast, should not be "blinded" by extremely weak radio noise at particular frequencies. This difference is part of what has made the quantum proposal feel testable rather than mystical. The same physics that makes singlet and triplet a tug-of-war also makes them, under the wrong electromagnetic conditions, easy to perturb.

Still, a caution is needed when words like "quantum" enter popular science. The radical-pair mechanism is quantum in the plain, unromantic sense that the relevant variables are quantum states of electron spins. It does not require consciousness, exotic entanglement across macroscopic distances, or any other fashionable flourish. The radical pair begins in a correlated state because it is created by a single electron transfer, and the subsequent evolution follows from standard quantum mechanics applied to two electrons coupled to nearby nuclei. That is all. The wonder is not that quantum rules exist, but that evolution might have found a way to exploit a microsecond-long quantum sensitivity in the service of a bird's kilometre-scale navigation.

A second caution concerns the word "direction". Even a perfectly functioning chemical compass would not tell an animal where it is. It would provide a heading, not a map. Anyone who has hiked with a compass knows the difference. A compass can tell you which way you are facing, but it cannot tell you which valley you are in. Many animals appear to possess both a magnetic compass and, in some cases, a magnetic map sense based on field intensity or inclination gradients. The radical-pair mechanism, as usually discussed for birds, is a candidate primarily for the compass component: a directional reference that can be integrated with stars, landmarks, wind, and smell. That division of labour helps explain why different magnetic mechanisms might coexist in the same animal.

The most persuasive way to hold this in mind is to return to the retina, not as a poetic location but as an engineered surface. The retina does not merely detect light; it extracts patterns. It compares neighbouring signals, enhances contrast, and compresses information before sending it down the optic nerve. A weak, direction-dependent modulation from cryptochrome chemistry would not have to be strong in absolute terms. It would have to be consistent, structured, and correlated across many receptors. The brain is extraordinarily good at extracting consistent structure from noisy sensory streams; that is what perception is. If cryptochrome outputs are biased by the geomagnetic field in a way that depends on molecular orientation, then the retina is the right kind of tissue to transform that bias into a usable cue.

There is a final subtlety that makes the tug-of-war analogy more than a storytelling device. In many chemical reactions, the initial conditions are forgotten almost immediately. Heat randomises motions, collisions scramble phases, and products depend mainly on average energies. Radical pairs are different because the initial spin state matters, and it matters *because* it restricts which reactions are allowed. The reaction is not free to explore all possibilities; it is channelled by spin selection rules. That constraint is precisely what allows a weak field to have an effect. The field is not battling against a thermal hurricane to push a reaction uphill. It is adjusting the timing of a quantum constraint that decides whether recombination is permitted at a given moment. When the constraint is the gate, the field only needs to jiggle the latch.

For a bird in migration season, the stakes of this microscopic contest are not microscopic at all. A few degrees of error, repeated night after night, can determine whether a bird reaches a safe wintering ground or ends up exhausted over the ocean, or in a landscape where food is scarce. The chemistry does not contain a map of Europe or a sense of destiny. It contains a rapid, fragile, exquisitely tuned competition between two quantum states, and that competition can be biased by the geometry of a planet's field. In the dark, inside a living eye, electrons keep score.

## 3.3 Blinded by the Buzz

A compass that can be knocked off course by a whisper of radio noise sounds, at first, like a plot device. Radios are everywhere; birds have been migrating for millions of years; surely a magnetic sense should be robust enough to ignore the electromagnetic clutter of modern life. And yet, in the laboratory, something close to that plot device has played out: under conditions where the Earth's field is carefully controlled and everything else is held steady, weak radio-frequency fields can make a migratory bird behave as if its magnetic compass has gone missing.

The drama matters because it is oddly specific. If an animal's magnetic sense were based on tiny magnets tugging mechanically on a receptor—magnetite grains acting like microscopic compass needles—then radio noise at ordinary intensities should do almost nothing. Oscillating fields at tens of megahertz change direction far too quickly for a physical particle embedded in tissue to follow, and the forces are vanishingly small. But if the sensor is chemical, and if its key intermediate is a radical pair whose behaviour depends on electron spins, then radio-frequency fields are exactly the kind of "gentle interference" that can scramble the signal. The difference is not philosophical. It is engineering: mechanical sensors respond to force and torque; radical-pair sensors respond to spin dynamics.

To see why, it helps to recall the core physical ingredient without rehearing the full story. A radical pair is two molecules, each with an unpaired electron, created together by an electron transfer. Those two electron spins begin in a linked quantum state and then interconvert between singlet-like and triplet-like character. The Earth's magnetic field subtly alters the tempo of that interconversion and therefore the chemical products. A bird does not need the Earth's field to be strong; it needs the field to be *predictable* so that the chemistry can be biased in a direction-dependent way.

Radio-frequency (RF) fields are time-varying magnetic fields. A broadcast at, say, 80 MHz means the field oscillates 80 million times per second. If the radical pair's internal spin dynamics naturally happen on microsecond timescales—fast enough to notice the Earth's field—then an RF field in the right frequency range can act like a mischievous metronome, nudging the spins out of their carefully constrained evolution. In the language of spin chemistry, RF fields can drive transitions or induce decoherence, disrupting the delicate singlet–triplet balance that the Earth's static field would otherwise bias.

There is an everyday analogy that comes close, if kept on a short leash. Imagine trying to keep time with a pendulum clock. The clock is not powered by a strong force; it is powered by a steady one, and its reliability comes from the regularity of the pendulum's swing. Now imagine someone lightly tapping the clock case at just the wrong rhythm. The taps do not have to be strong enough to knock the clock off the wall. They only have to spoil the regularity. The clock still swings, but it stops telling the truth. In a radical-pair compass, RF fields are those taps. They do not need to overwhelm the reaction; they need to disrupt the timing.

This possibility moved from theory to unsettling behavioural evidence in a study by Svenja Engels and colleagues (2014). The work was striking not only for its result but for its setting. Many laboratory studies of bird orientation take place in highly controlled rooms, far from the electromagnetic mess of a city. Engels and her collaborators designed an experiment that treated the city as part of the question. They tested European robins in huts that differed mainly in their shielding: some were electromagnetically screened and grounded, others were not. In the screened huts, robins oriented as expected using the magnetic field. In the unscreened huts, exposed to the background electromagnetic noise of the environment, the birds became disoriented: their headings scattered rather than clustering around a meaningful direction. When broadband noise was deliberately introduced into the screened huts, the disorientation returned.

There is a particular kind of chill that comes from such an experiment, because it suggests a new category of pollution. Air pollution burns lungs; light pollution hides stars. Electromagnetic noise, in this picture, might scramble a sensory channel that was never designed to cope with the electronic exuberance of modern infrastructure. The robins in the study were not harmed in any obvious way; they simply lost a compass cue under certain electromagnetic conditions. For a bird attempting a nocturnal migration across unfamiliar terrain, a temporary loss of magnetic orientation could mean wasted energy, delayed arrival, and increased mortality.

At the same time, the result demands caution. Behavioural experiments are vulnerable to confounds, especially with animals as sensitive as birds. Temperature gradients, tiny drafts, subtle sounds, vibrations, stress, and unrecognised light cues can all influence orientation. Shielding a hut changes more than radio fields; it can alter acoustics and microclimate. The strength of the 2014 study was its careful attempt to isolate the electromagnetic variable, including the "restore and disrupt again" logic that makes alternative explanations less likely. But the broader literature has not always been uniform across species, labs, and conditions. Some RF effects replicate readily; others are capricious. That is not surprising if the underlying sensor is a fragile, context-dependent chemical process embedded in living tissue.

The most persuasive RF experiments are those that behave like a spectroscopic probe: not merely "noise ruins it," but "particular frequencies and bandwidths ruin it in a way that constrains the underlying chemistry." This is where more recent work becomes scientifically rich. A study in 2022 reported that weak broadband RF fields in the 75 to 85 MHz range could disrupt magnetic compass orientation in night-migratory songbirds, and the authors connected the frequency dependence to specific features of flavin-based radical pairs. The implication is not that birds have an "80 MHz antenna." The implication is that the relevant spin system has characteristic energy splittings and hyperfine couplings that make it especially susceptible to perturbation in certain bands. Frequency dependence, in other words, can turn behavioural disruption into a clue about which radicals are involved.

That phrasing also reveals a central uncertainty. Even if cryptochrome is the receptor scaffold, which precise radical pair does the sensing? In cryptochromes, electron transfer often involves a flavin (FAD) and a chain of tryptophans, producing radical pairs such as $FAD{\cdot}^{-}$ with $Trp{\cdot}^{+}$. But alternative partners are possible, and the photochemical cycle includes multiple redox states and potential radical intermediates. Behavioural timing experiments, including cases where magnetic orientation seems possible during dark intervals after light exposure, hint that the critical radical pair may form during a later step in the cycle, perhaps during re-oxidation. If so, the radicals doing the sensing might not be the ones most obvious in a simplified textbook diagram.

This is where the RF story and the biochemical story strengthen one another. Biochemistry has recently provided unusually direct evidence that cryptochrome 4 from a migratory songbird can show magnetic-field-dependent reaction yields in vitro, and that mutating elements of the electron-transfer chain alters that sensitivity. Those results do not automatically prove that the same radical pair is responsible for the bird's behavioural compass in vivo, but they make it much less plausible that the compass is *entirely* unrelated to cryptochrome radical chemistry. When behavioural RF disruption is placed alongside in-vitro magnetosensitivity in a candidate molecule, the two lines of evidence feel less like coincidences and more like different faces of the same mechanism.

Still, one should resist the temptation to treat RF disruption as a flawless "quantum litmus test." There are at least three reasons.

First, RF environments are complicated. A city is not a single frequency; it is a stew of signals from power infrastructure, communication systems, electronic devices, and the unintended emissions of poorly shielded electronics. Measuring what a bird is actually exposed to at the location of its head, inside a cage, is not trivial. A weak RF field on a meter may translate into a different pattern near metal supports, cables, or grounded meshes.

Second, biology is not a uniform detector. The radical-pair mechanism predicts susceptibility to certain oscillating fields, but the magnitude of susceptibility depends on coherence times, relaxation rates, molecular orientation, and the way the chemical output is transduced into neural signals. Two species might share the same basic mechanism yet differ in vulnerability, simply because one species' receptor complex is better shielded in tissue, more strongly ordered in the retina, or coupled to downstream signalling in a way that averages over noise.

Third, there may be more than one magnetic mechanism in play. Many animals likely use magnetite-based receptors for some magnetic tasks, especially those involving field intensity or map-like information. A bird might possess both a chemical compass for direction and a mineral-based sensor for other purposes, or it might shift strategies depending on lighting conditions and ecological context. In such a mixed system, RF disruption of the radical-pair compass might not always translate into obvious disorientation if other cues can compensate.

Even with these caveats, the behavioural phenomenon is hard to ignore: under some conditions, weak RF fields can erase the directional behaviour that otherwise tracks a magnetic cue. That is a rare gift in sensory biology. Most senses are hard to disrupt selectively; block the eyes and you change much more than light input, because you change stress, posture, and behaviour. Here, one can apply an invisible physical stimulus that—in principle—targets a very specific molecular degree of freedom: electron spin. The possibility of reaching into an animal's sensory system at that level feels almost indecently precise.

A useful way to visualise what "disoriented" means in these experiments is to imagine a circular histogram of the directions a bird attempts to take. If the bird has a stable compass heading, the data cluster. If the compass is scrambled, headings scatter around the circle. The plot below is an illustrative version of the kind of figure often used in the field, showing the contrast between directional behaviour under shielded conditions and near-random headings under RF exposure. The numbers are synthetic, meant to convey the idea rather than reproduce a particular dataset.

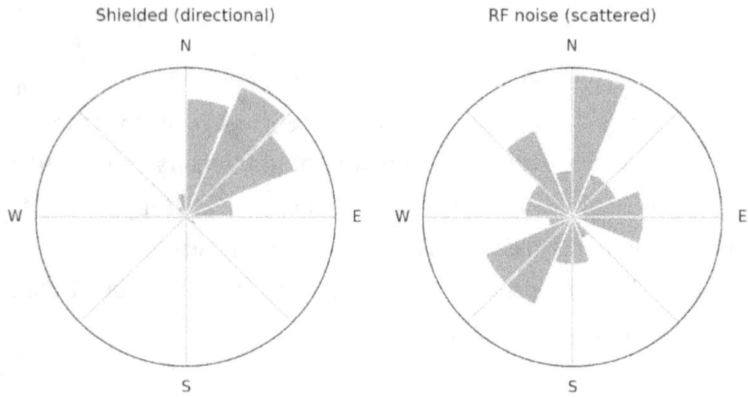


The deeper meaning of these plots is not simply that birds can be confused. It is that the confusion is triggered by a stimulus that plausibly targets the mechanism's most distinctive vulnerability. If the compass depends on coherent spin dynamics, then RF noise is not just "another disturbance." It is interference aimed at the exact degree of freedom that makes the compass possible.

This brings the discussion back to a human-scale concern. Modern electromagnetic environments are not uniform. A forest clearing is not a rooftop. A rural valley is not an airport perimeter. If RF noise can disrupt magnetic orientation under some intensities and bandwidths, then different landscapes could become unevenly navigable in ways that are invisible to us. A bird might pass through a corridor of electromagnetic quiet where its magnetic compass works normally, then cross a patch of electromagnetic clutter where the compass signal collapses and must be replaced by stars, landmarks, wind, or sheer statistical correction over time. The bird's behaviour could still look competent in the long run, but with hidden costs: longer routes, more stops, greater vulnerability in bad weather.

None of this requires panic, and none of it implies that migration is about to fail wholesale because of radio transmitters. Animals rarely

rely on a single cue. The magnetic compass is one component in a toolkit that includes celestial information, polarized light patterns, odours, topography, and learned landmarks. But it does suggest that when researchers argue over whether the magnetic compass is "really quantum," they are not merely playing with words. They are pointing to a mechanism that has a very particular Achilles' heel, and that Achilles' heel has become a feature of the modern world.

There is a final irony. The same electromagnetic technology that may interfere with a bird's compass is also what allows humans to probe the compass scientifically. By generating controlled RF fields, researchers can perturb the proposed mechanism gently, reversibly, and in a frequency-specific way. It is a rare situation in biology where the suspected molecular mechanism comes with its own experimental dial. Turn the dial and the animal's behaviour changes. Turn it back and the behaviour returns. Somewhere between those two states lies a radical pair, trying to keep its quantum balance in a noisy world.

## Chapter 4

# Living Magnets and Iron Nerves

*While quantum mechanics operates in the retina, a heavier, more ancient tool lies hidden in the dark recesses of the beak and nose: biogenic magnetite. This chapter examines how evolution synthesizes literal lodestones within cells to create a sensor capable of measuring magnetic intensity. We uncover the physical machinery that turns the Earth's field into a mechanical tug, providing the essential "map" that tells an animal not just which way is North, but exactly where it stands.*

## 4.1 The Living Lodestone

A lodestone is a rock that behaves as if it has a will of its own: bring it near iron filings and they leap up, arranging themselves into delicate spines that seem to reach for the stone. Long before anyone spoke of electrons or domains, this small drama suggested a particular kind of causation. The stone does not *persuade* the filings chemically, the way vinegar persuades chalk to fizz. It *pulls* them, at a distance, with a force that appears to pass through intervening air as if air were not there at all.

That distinction matters because one route to magnetic sensing in animals is, at heart, lodestone-like. It is mechanical. It turns invisible geometry into physical tension. Somewhere in a living tissue, a microscopic magnet behaves like a needle in a compass, and its attempt to align with the Earth's magnetic field can be converted into the opening of an ion channel: a gate in a cell membrane that lets charged particles flow. A trickle of ions becomes an electrical signal; an electrical signal becomes a sensation.

The main actor in this story is magnetite, an iron oxide with the formula $Fe_3O_4$. Magnetite is not the only iron mineral in biology, and in many animals it is not even the most abundant form of stored iron. Ferritin, for example, is a protein cage used to keep iron safely locked away. But ferritin is not a bar magnet. Magnetite can be. Under the right conditions it forms crystals that carry a stable magnetic moment, meaning they behave like a tiny permanent magnet with a north and south pole.

It is easy to imagine magnetism as a property that appears only when a substance is macroscopic: a horseshoe magnet on a school desk, the magnets that hold photographs to a refrigerator. At the scale of molecules, so the intuition goes, everything must jiggle too much for a magnetic direction to remain meaningful. Biology complicates that intuition. Some organisms routinely manufacture magnetite crystals that are neither random grains nor flimsy clusters, but neat, uniform structures whose size and shape are tuned to magnetic physics.

A clear historical turning point came in 1975, when Richard Blakemore described bacteria that swim persistently along magnetic field lines. These *magnetotactic* bacteria contain chains of magnetite crystals, each one enclosed in a membrane and arranged like beads on a string. In a droplet of water the organisms behave as though they carry a compass inside their bodies. The elegance of the solution is almost unsettling: instead of a complicated chemical computation, the bacterium simply builds a physical needle and lets the planet do the orienting. The bacterial example does not prove that birds, turtles, or salmon use the same mechanism, but it proves something more basic and more important: evolution can build an intracellular magnet that is strong enough to matter.

To understand why size matters, it helps to introduce one term from magnetism that sounds forbidding but describes a simple idea. A crystal of magnetite can be *single-domain*. In a large magnet, different regions (domains) can point their internal magnetization in different directions; the domains rearrange, and the magnet can be demagnetized or remagnetized. In a sufficiently small crystal, domains are energetically unfavorable, so the entire crystal behaves as one coherent magnetic unit. It is a single tiny bar magnet. This is where biology often aims. If the crystal is too large, multiple domains can form and the net moment may wobble or vanish. If it is too small, thermal agitation can overwhelm magnetic stability and the moment flips randomly (a superparamagnetic regime). Between these extremes lies a sweet spot where the moment is stable but the object is still small enough to be built, positioned, and coupled to cellular machinery.

The Earth's magnetic field is weak by everyday standards. A typical refrigerator magnet produces a field many thousands of times stronger at its surface than the field that threads through a bird's head. The Earth's field near the surface is on the order of tens of microtesla, varying by location. The puzzle is not that a magnetite particle *can* feel such a field; the puzzle is whether the field can exert a force or torque that stands out above biological noise.

The relevant quantity is the magnetic moment $m$ of the particle, which is roughly its magnetization times its volume. The field **B** exerts a torque

$$\boldsymbol{\tau} = \mathbf{m} \times \mathbf{B},$$

tending to rotate the particle so that **m** aligns with **B**. The associated energy difference between aligned and anti-aligned states is on the order of

$$\Delta E \sim 2mB,$$

and whether alignment is meaningful depends on how this compares with the thermal energy scale $kT$ (Boltzmann's constant times temperature). At body temperature, $kT$ is about $4 \times 10^{-21}$ J. A single-domain magnetite crystal with dimensions of a few tens of nanometres can have a moment large enough that $mB$ becomes comparable to, or exceeds, $kT$ in the Earth's field. That does not mean the particle locks rigidly into place like a compass needle on a pin. It means the particle's orientation is biased by the field: it spends more time near alignment than it would by chance. Bias can be enough if the cell has a way to read it out.

Cells are full of ways to read tension. Ion channels can be mechanosensitive: they change their probability of opening when the membrane is stretched or when a tether pulls on part of the protein. In the ear, mechanically gated channels in hair cells are responsible for the first step in hearing. In touch receptors in skin, similar principles operate. A magnetite-based sensor borrows that logic. Instead of sound vibrating the hair bundle, the Earth's field biases the orientation of a tiny magnet. Instead of a fingertip deforming, the magnet's rotation tugs on a molecular linkage.

A single isolated crystal, floating freely in cytoplasm, would not help much. It would simply rotate with the surrounding fluid, and any torque would be dissipated. A useful receptor requires *coupling*: the magnet must be linked to something that can translate its rotation into a local mechanical change. Bacteria solve this by arranging magnetite crystals into chains anchored within the cell. A chain behaves differently from a lone particle. It has a larger net moment, for one thing, since the moments add. It also resists random rotational diffusion more strongly because it is an elongated object, and its magnetic anisotropy (a preference for magnetization along a particular axis) is reinforced by shape.

The chain is not simply a pile of grit. In magnetotactic bacteria it is constructed with remarkable precision. The crystals are often nearly identical in size and shape, and they are held at fairly regular spacing by a cytoskeletal scaffold. The result is a nanoscale compass needle, engineered by proteins rather than by human hands. In that light, the idea that larger animals might also build intracellular magnets is less fanciful than it first sounds. The core requirement is not intelligence but evolutionary opportunity: a way for small mutations in iron handling, crystal nucleation, and cellular scaffolding to be selected because they improve orientation and navigation.

Now consider what a mechanosensitive transducer might look like in an animal cell. One plausible arrangement places a magnetite chain near the membrane, with one end coupled to cytoskeletal elements and the other end linked, directly or indirectly, to an ion channel. The membrane itself is not a rigid wall; it is a thin, flexible bilayer with embedded proteins. Pulling on a channel or altering local membrane tension can change the channel's conformation. A torque on the chain can therefore become an electrical signal.

The geometry matters. Torque is directional: it tries to rotate the magnet toward alignment with **B**. If the chain is constrained so it cannot freely spin, the torque becomes strain in its attachments. That strain can be tiny in absolute terms and still be decisive at the scale of proteins, where forces of a few piconewtons can alter conformation. Biology routinely uses such forces. Molecular motors like myosin and kinesin work in this force range; so do many mechanosensitive channels.

One can make the picture more concrete with a thought experiment grounded in ordinary experience. Hold a compass close to a strong magnet and you can see the needle twist until it points not north but toward the nearby magnet. Now replace your fingers with molecular tethers. Replace the compass needle with a chain of nanocrystals. Replace the visible motion with a subtle shift in the probability that a channel opens. The same physics is doing the work, only the readout has changed.

There is also an appealing way to explain why an animal might build *many* such units rather than a single large magnet. If one receptor were responsible for all magnetic information, it would be dangerously vulnerable: a small injury, a local infection, a developmental glitch. Arrays of receptors can average noise, detect gradients, and provide redundancy. In engineering, a set of small sensors often outperforms one large one, because it can encode spatial patterns. Magnetite-based mechanisms lend themselves to that strategy because each unit is compact.

The strongest argument in favor of magnetite-based reception has always been physical plausibility. Unlike mechanisms that rely on extremely subtle changes in chemical reaction rates, a magnetite particle in a magnetic field is subjected to a straightforward, classical torque. The Earth supplies a stable directional reference. The remaining question is biological: where are the crystals, how are they arranged, and how is their motion coupled to the nervous system?

Those biological questions are unusually hard because iron is everywhere. Blood is iron-rich. Many tissues contain iron-handling proteins. Even dust contamination can mislead. Stains that detect iron can light up cells that are simply storing or processing iron rather than sensing magnetism. A magnetoreceptor, if it exists, must be embedded within the ordinary logistics of iron metabolism, and that makes it difficult to pick out anatomically. It is like trying to find a particular musician in a city where everyone owns an instrument.

Still, there are recurring motifs that keep returning whenever magnetite is discussed seriously. One is the insistence on *stereotypy*: a true receptor should be located in consistent places across individuals of a species, with reproducible structure and neural connections. Another motif is *association with nerves*. A magnetite particle by itself is not a sense. A sense requires transduction and transmission. If magnetite is doing the transduction, it should be near cellular machinery capable of producing a receptor potential and near nerve fibers capable of carrying that signal.

A third motif is the *difference* between a compass and a map. A compass sense answers: which way should I go? A map-like sense answers: where am I? Magnetite mechanisms are naturally suited to a map component because the Earth's field varies in intensity and inclination across geography. A receptor that responds to intensity could provide a cue that changes gradually along a migratory route. That is not the same as direction, and it is worth holding the distinction in mind because magnetite-based transduction is often discussed alongside a very different, light-dependent magnetic mechanism. The two can coexist without redundancy: one provides direction, the other provides position-related information.

The most vivid experimental signature of magnetite is not found in a microscope but in behaviour under magnetic pulses. Ferromagnetic materials like magnetite can be remagnetized by brief exposure to a strong field. If an animal's magnetic sensitivity depends on a stable magnetic remanence, then a pulse could, in principle, flip or perturb the internal magnet and produce predictable errors. This idea has driven decades of experiments, and it has a particular charm because it treats the animal not as a mysterious oracle but as a piece of physical apparatus: apply a pulse, observe the change, infer the material.

Yet even here, the story resists being too neat. A pulse does not act on all magnetic materials equally, and an animal is not a rigid compass. If magnetite crystals are arranged in multiple orientations, or if they are coupled to elastic structures that relax, or if the animal integrates magnetic information with celestial cues, the behavioural outcome may be subtle. The same pulse might cause a dramatic change in one context and almost nothing in another. That variability is frustrating, but it is also informative: it suggests the magnetite system, if present, is embedded in a larger sensory ecology rather than operating as a solitary instrument.

There is also a common misconception worth clearing away. Calling the mechanism "magnetite-based" does not mean the animal is carrying a chunk of iron that points like a compass needle visible in a cartoon. The likely scale is nanometres to micrometres. The motion is not a sweeping rotation but a slight bias in orientation, perhaps a fraction of a degree, translated into changes in channel gating. The sensor does not need to swing freely; in many models, it must *not* swing freely, because the signal is strain, not motion.

One can also be precise about what kind of signal such a sensor could plausibly deliver. A magnetite-based mechanoreceptor would be sensitive to the *vector* nature of the field: direction matters because torque depends on the angle between **m** and **B**. If the chain is anchored in a fixed orientation in the body, then the torque will change as the animal turns. That provides a directional cue. If, in addition, the receptor's response depends on field strength $B$ (because torque magnitude is $mB \sin\theta$), then changes in intensity can modulate the signal. In principle, a population of receptors with different orientations could encode both.

The dream scenario for a physiologist would be to isolate one receptor cell, record its membrane potential, and watch it respond as the field is rotated or intensified under controlled conditions. Such recordings are difficult in animals for all the reasons already mentioned: rarity, uncertain location, and the risk of artefact. But biology has a habit of using the same design logic in different places. Mechanically gated channels are real and common. Cytoskeletal tethers are real and common. Intracellular crystals are real and, in some organisms, exquisitely controlled. From those ingredients, the idea of a living lodestone is not a metaphor but a plausible piece of cellular engineering.

The point can be appreciated in a final, tactile image. Imagine a door latch held shut by a spring. You can open it by pushing chemically, perhaps by dissolving the latch, but you can also open it by pulling on a string. Magnetite provides the pull. The Earth provides the direction of pull. The cell provides the latch.

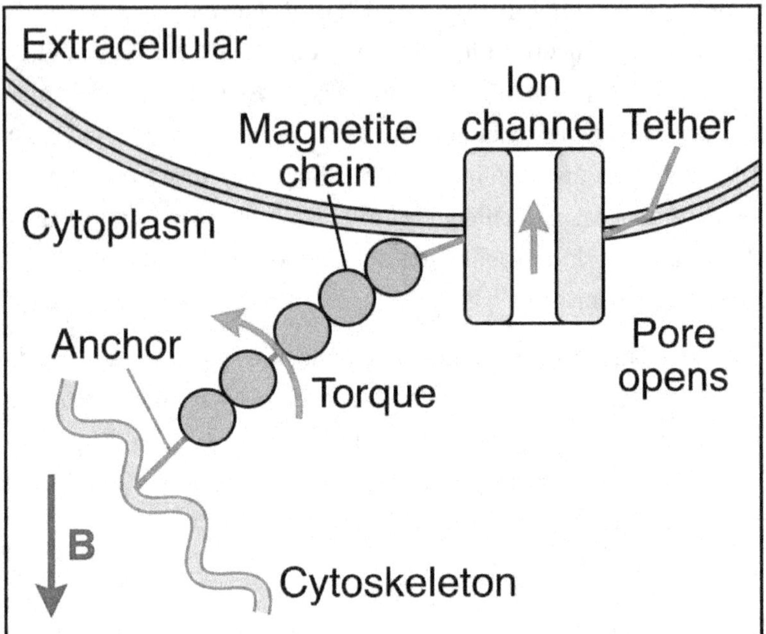


Rotation biases tension on the tether, shifting channel gating probability.

A diagram like this inevitably simplifies. Real membranes ripple; proteins flex; crystals may not be perfect spheres; the coupling could involve accessory proteins rather than a single "tether." But the core idea is stubbornly concrete. A magnetic field can do mechanical work on a magnet. Mechanical work can modulate an ion channel. An ion channel can start a nerve signal. Somewhere between the elegance of a compass needle and the messiness of living tissue, magnetite offers a way for geography to reach directly into physiology.

## 4.2 Shocking the System

A compass needle has a weakness that can be turned into a test. Bring a strong magnet close and the needle does not merely wobble; it can be coerced into pointing somewhere else. Pull the magnet away and, depending on the needle and the treatment, it may return obediently to north or it may retain a trace of its miseducation. Anyone who has ever played with a needle and a bar magnet has performed, in miniature, an experiment that sensory biologists later tried to perform on animals: if the animal carries an internal piece of ferromagnetic material, can a brief blast of magnetism *reset* it?

The appeal of the idea is obvious. Many proposed mechanisms for magnetic sensing are subtle enough to make a sceptic restless. If the sense depends on delicate chemistry or on indirect effects in the nervous system, then an experiment that behaves like a workshop demonstration has a special power. Magnetite, unlike most biological materials, is ferromagnetic: it can hold a persistent magnetization. Ferromagnetism is why a refrigerator magnet remains a magnet when you remove it from the factory's electromagnets. It is also why a magnetite crystal, once magnetized, can behave as a tiny permanent magnet rather than a particle that only responds while a field is present.

A magnetic pulse experiment treats the animal's body as the place where a refrigerator magnet might be hiding. The animal is exposed, very briefly, to a field far stronger than the Earth's field. The pulse is usually too short to heat tissue significantly, but strong enough to act on magnetic minerals. Then one watches what changes. Does the animal's orientation shift by a predictable angle? Does its ability to discriminate magnetic conditions collapse? Does a previously learned magnetic association become scrambled? If so, it becomes difficult to avoid the conclusion that some part of the sensory apparatus contains a ferromagnetic element whose internal state can be altered.

To make sense of what a pulse can do, it helps to borrow two ideas from basic magnetism: *remanence* and *hysteresis*. Remanence is the magnetization that remains when an external field is removed. Hysteresis is the stubbornness of a ferromagnet: the material's magnetization does not depend only on the field it feels right now, but also on its past. When you sweep a field up and down, the magnetization follows a loop rather than a single line. In practical terms, hysteresis means you can "write" magnetic states into a particle, flipping its internal alignment if the applied field exceeds a threshold called the coercive field.

A magnetite-based receptor can be imagined in at least two broad ways. In one picture, a chain of single-domain crystals is already a stable magnetic dipole; the Earth's field applies a torque, and that torque is coupled mechanically to ion channels. In another, the receptor uses changes in magnetization itself (for example, shifts between two stable states) to create forces on nearby structures. Either way, a strong pulse is not just an exaggerated version of the Earth's field. It can change the internal magnetic configuration: flipping polarity, partially demagnetizing, or, if the pulse is extreme relative to the particles' stability, disrupting the arrangement enough that the mechanical linkage no longer behaves as it did. A pulse is a way of reaching into the receptor and turning a knob that the environment cannot normally turn.

The elegance of pulse experiments is matched by their hazards. A ferromagnet is not the only thing that can be influenced by a strong, rapidly changing magnetic field. Conductive loops can experience induced currents. Nervous tissue is electrical. If a pulse produces a behavioural change, one must rule out a nonspecific jolt to the nervous system rather than a specific reprogramming of magnetic particles. This is why the most persuasive pulse studies are careful about timing, controls, and the particular pattern of behavioural errors they produce. A general stress response tends to make animals less oriented; a specific perturbation of a magnetic sensor tends to make them oriented *wrong* in a repeatable way.

One of the most discussed pulse phenomena is a systematic shift in migratory headings. Migratory birds tested in orientation funnels often show a clear preferred direction, as if the night sky were compelling them to aim their restlessness. In some species and conditions, applying a strong magnetic pulse changes that preferred direction by a predictable amount. The pulse does not usually erase orientation altogether; instead it produces a rotated or otherwise altered preference, which is precisely what one would expect if the compass information has been biased or recalibrated.

Wolfgang and Roswitha Wiltschko's group has been central to developing the logic of these experiments, particularly the idea that pulse effects point toward magnetite. In work on migratory silvereyes reported by Wiltschko, Munro and colleagues (2009), experienced birds showed a striking post-pulse change in their migratory direction, roughly a right-angle shift under the test conditions. The critical move came when the researchers locally anesthetized the skin of the upper beak. With the beak region temporarily deactivated, the same pulse no longer produced the characteristic shift. A magnetite-based system located in or near the beak, connected to the nervous system via sensory fibres, offered a neat interpretation: the pulse "reset" a magnetic component, and numbing the beak removed the route by which that reset could influence behaviour.

There is a reason this sort of result lands so forcefully when people first hear it. It does not rely on a microscope image that might be artefactual, or a staining pattern that might be misread. It relies on an intervention that is almost insultingly direct. If you can switch off a pulse effect by numbing a small patch of tissue, you have constrained the anatomy. If you can create the effect with a physical treatment that ferromagnets are known to respond to, you have constrained the physics.

And yet the field learned, sometimes painfully, that even an experiment that looks like a clean "smoking gun" is not immune to ambiguity. Not every species shows pulse effects, and not every behavioural paradigm reveals them. Even within birds, pulse effects can depend on age, migratory state, or the cues available during testing. That variability is not merely an annoyance; it is a clue that magnetic information is being integrated with other senses, and that the part of the magnetic system affected by pulses may not always be the part dominating the animal's decisions.

There is a second line of work that makes the pulse logic feel less like a parlour trick and more like sensory physiology: conditioning animals to detect magnetic anomalies. Cordula Mora and Michael Davison (2004) designed experiments in which pigeons learned to choose between alternatives based on whether a magnetic anomaly was present. The task turns magnetoreception into something closer to hearing or vision: a discriminable stimulus that can be associated with reward. When the upper beak region was locally anesthetized, performance suffered. When the ophthalmic branch of the trigeminal nerve (often called V1), which innervates parts of the upper beak and face, was cut on both sides, performance also suffered. These results did not, by themselves, prove magnetite, but they gave pulse experiments a natural home. If a beak-associated pathway conveys magnetic information, then a pulse that perturbs magnetite-like material in that pathway could plausibly distort the signal.

Pulse experiments also carry a particular kind of physical signature that is hard to imitate with purely chemical mechanisms. A chemically based magnetic effect, such as one involving radical pairs, is typically sensitive to weak oscillating radio-frequency fields and depends on the direction of the magnetic field relative to molecular axes, often in a light-dependent way. A brief strong pulse, by contrast, is most naturally suited to reorienting ferromagnetic particles with hysteresis. That does not mean pulses cannot affect other systems, but it does mean that when a pulse produces a long-lasting change in magnetic behaviour, magnetite becomes an obvious suspect.

To see what "long-lasting" means here, imagine two compasses. One is an ordinary needle on a pin. Bring a strong magnet near and the needle swings, but when you remove the magnet the needle returns to pointing north. The second compass is a cheap souvenir with a magnetized strip that can be partially demagnetized; expose it to a strong field and it may remain biased afterward. The difference is not the strength of the field at the moment of exposure but the presence of remanence: a memory in the material itself. Many pulse experiments are looking for exactly this kind of memory. If the animal's behaviour is altered for minutes, hours, or longer, something in the sensory chain has retained a changed state.

A simple cartoon of the hypothesis would be: pulse flips the polarity of a magnetite chain; the chain now exerts torque in the opposite sense for a given body orientation; mechanosensitive channels are gated differently; the nervous system reads a systematically distorted signal; the animal's chosen heading shifts. Real receptors are unlikely to be so binary, but the logic remains: a pulse does not merely add noise; it changes the baseline.

A good experimental logic also asks for reversibility and specificity. If a pulse can shift orientation, can a pulse in the opposite direction shift it back? If the effect depends on a particular part of the body, does shielding or numbing that region reduce it? If the animal is tested in a context where magnetic cues are irrelevant (for example, with strong visual landmarks dominating), does the pulse still produce confusion? These are the kinds of questions that make the difference between a dramatic effect and a persuasive one.

At this point, a complication enters that is as sociological as it is scientific. Magnetite is a politically charged molecule in magnetoreception research because it seems to promise a straightforward mechanism. Some researchers distrust it for that very reason: nature is rarely so obliging. Others embrace it because it anchors the sense in classical physics rather than in delicate quantum chemistry. Pulse experiments have therefore carried a symbolic weight that sometimes exceeds their technical details. A positive pulse effect is taken as vindication by one camp and as a challenge to be explained away by another.

The healthiest way to treat the evidence is to let it remain stubbornly mixed. There are pulse studies that show striking, repeatable behavioural changes consistent with ferromagnetic involvement, and there are studies that do not. GPS tracking of homing pigeons in at least one reported case found no measurable influence of pulsing on routes or homing performance, a result that is difficult to reconcile with the idea of a single, simple compass needle that dominates all magnetic behaviour in all birds. Such outcomes do not demolish the magnetite hypothesis; they do, however, insist that the magnetic sense is not a single instrument that can be knocked out with one blow. It is more likely a collection of channels whose importance depends on context: migration versus homing, familiar terrain versus unfamiliar, clear skies versus overcast, and perhaps even the internal state of the animal.

The mismatch between some strong pulse effects and some null results also forces a more careful thought about what, exactly, a pulse might be perturbing. If magnetite-based receptors contribute mainly to a *map*-like signal based on intensity, then a pulse might scramble position estimates without necessarily destroying the ability to keep a steady compass direction. In a homing task where landmarks and olfactory cues are rich, the animal might compensate. In a tightly controlled orientation test where the decision hinges on magnetic input, the perturbation might be exposed. The same physical intervention could therefore look decisive in one paradigm and invisible in another, not because one study is sloppy but because the animal's brain is not obliged to use magnetic information in the same way every time.

There is also the more unsettling possibility that a pulse effect reveals something real about magnetic material in the body without revealing a *receptor*. Many tissues contain iron compounds for mundane reasons. If a pulse acts on those compounds and indirectly changes physiology, behaviour could shift without any magnetic sense being involved. This is why the best pulse work links the effect to specific sensory pathways, such as the trigeminal system, and why combining pulse interventions with localized anesthesia or nerve manipulations has been so influential. The closer one can get to the anatomy of sensation, the less room there is for a general "magnetic shock" explanation.

Despite the controversies, pulse experiments have done something valuable for the field beyond any single result: they have forced magnetoreception to become falsifiable in a mechanistic way. If a proposed receptor contains ferromagnetic material with remanence, then there are predictions about how it should behave under pulses of different strengths and directions. If the receptor is purely chemical and depends on weak-field effects, then there are different predictions. In other sensory systems, this kind of physical probing is routine. Vision was not understood by staring at eyes; it was understood by controlling light. Hearing was not understood by admiring ears; it was understood by delivering vibrations and measuring responses. Pulse experiments are an attempt to deliver magnetism with the same experimental confidence.

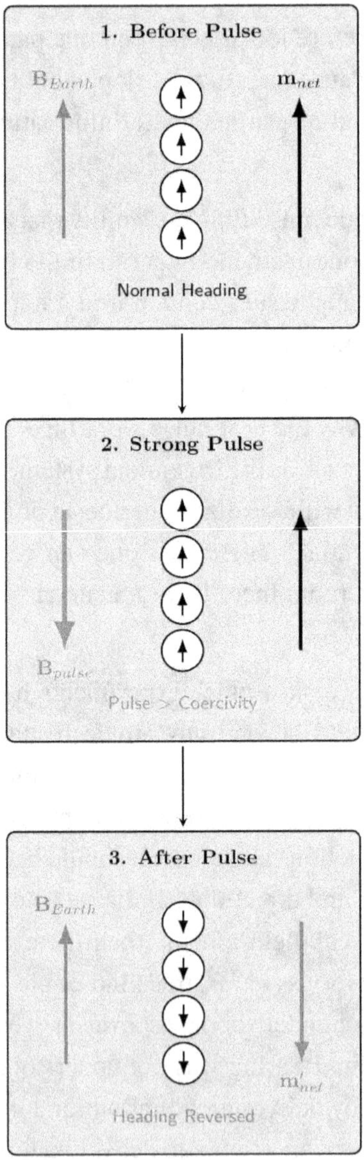

The diagram is deliberately modest: it does not insist that biology contains a literal bar magnet that flips neatly end-for-end. It merely captures the experimental ambition. A pulse is an attempt to force an internal magnetic element into a different state, and then to see whether the animal behaves as though one of its navigational rulers has been subtly re-graduated.

Perhaps the most human aspect of this work is the mixture of audacity and restraint it demands. Audacity, because it takes nerve to expose a living creature to a magnetic treatment strong enough to rewrite a ferromagnet and then claim that its subsequent choices are a window into invisible crystals. Restraint, because the temptation to declare victory is always premature. A pulse effect can be spectacular and still be misinterpreted; a null result can be real and still be uninformative if the animal had other cues to lean on.

What remains difficult to shake, though, is the basic asymmetry. A chemical compass can be disrupted by the kind of electromagnetic "static" that leaves a permanent magnet indifferent, while a ferromagnetic particle can be remagnetized by a pulse in a way that a fleeting chemical state cannot easily mimic. Pulse experiments exploit that asymmetry. They are not a proof on their own, but they are a way of asking nature a question in the blunt language of physics: if there is iron in the sense organ, will it remember being shocked?

## 4.3 The Trigeminal Map

A pigeon's beak looks like a simple tool: a hard, neat wedge for pecking grain, preening feathers, lifting twigs. But the beak is also a dense sensory surface. It is laced with touch receptors, temperature receptors, pain fibres, and the nerves that tell the brain whether a seed is cracked or still safely encased. For an animal that lives by pecking at the world, the beak is not an ornament; it is a hand.

That is why the most intriguing phrase in avian magnetoreception is not “compass” but “map”. A compass tells you which way to go; a map tells you where you are. For a bird released hundreds of kilometres from its loft, the practical question is not merely which direction is north. It is whether home lies upwind or downwind, over the next ridge or beyond the next sea. If the Earth’s magnetic field is part of that answer, it makes sense that the relevant information might arrive through a sensory channel that already specializes in local, bodily cues: the trigeminal nerve.

The trigeminal nerve is the great sensory highway of the face. In humans it carries touch and pain from the forehead, eyes, cheeks, and jaw, and it supplies the reflexes that make us blink when something approaches the cornea. In birds the same basic plan holds, with a branch that innervates the upper beak and regions around the nostrils. Anatomists call that branch the ophthalmic division, often abbreviated V1. The name sounds eye-centred, but in birds it has extensive territory in the beak and cere (the fleshy area at the base of the beak in pigeons). If a magnetic sensor sits in the front of the head, V1 is a plausible courier.

The idea that the beak might house a magnetic “map” grew from a particular kind of experiment: the sort that does not ask an animal which way it wants to go, but instead asks whether it can *detect* a magnetic condition at all. Orientation tests can be ambiguous because an animal can choose a direction for many reasons and can compensate with other cues. Detection tasks are blunter. They ask: can the animal tell the difference between magnetic situation A and magnetic situation B?

In the early 2000s, Cordula Mora and Michael Davison developed a conditioning paradigm with pigeons that made this question experimentally tractable. Pigeons were trained to make different choices depending on the presence or absence of a magnetic anomaly. The details are less important than the logic: if the pigeon is responding to magnetic information rather than to stray cues, then manipulations that disrupt magnetic input should impair performance.

Two manipulations did. One was local anesthesia applied to the upper beak region, temporarily silencing sensory endings there. The other was cutting the ophthalmic branch of the trigeminal nerve on both sides, preventing sensory signals from that region from reaching the brain. When these interventions reduced the birds' ability to perform the magnetic discrimination, it became harder to treat "magnetic map" as a purely cognitive trick or as a by-product of the visual system. The front of the head began to look less like a convenient place to put a mythical compass needle and more like an actual sensory organ with wiring.

Wiring matters because biology is not impressed by clever physical mechanisms unless they can be connected to behaviour. A chain of magnetite crystals, however elegant, is only a hypothesis until it can be placed into the circuit of perception: receptor cell, afferent nerve fibre, brainstem relay, higher processing. The trigeminal story is compelling precisely because it offers such a circuit, at least in outline.

The ophthalmic branch (V1) carries sensory axons from the beak region to the trigeminal ganglion, a cluster of neuronal cell bodies that sits near the skull base. From there, axons project into the brainstem, terminating in trigeminal nuclei that process facial sensation. Two regions often named in the magnetoreception literature are the principal sensory trigeminal nucleus (PrV) and parts of the spinal trigeminal nucleus (SpV). In an animal, these nuclei are not "magnetic centres" in the way the retina is a "visual centre". They are more like switchyards: places where sensory signals are sorted, amplified, inhibited, and routed onward.

One of the clever ways to test whether a pathway is carrying information is to use the brain's own bookkeeping. When neurons are strongly activated, they express certain genes rapidly, producing proteins that can be detected later by staining. These are called immediate early genes. One of the most frequently used markers in birds is ZENK. The name is not essential; what matters is that it provides a kind of metabolic footprint of neural activity.

Dominik Heyers, Manuela Zapka and collaborators (2010) used this approach in migratory birds by exposing them to magnetic conditions that changed over time and then examining the brainstem for increased ZENK expression. The result that drew attention was not a diffuse glow of activation everywhere, but a localized increase in trigeminal brainstem regions associated with trigeminal input. When the ophthalmic branch was cut on both sides, that magnetic-field-dependent activation largely disappeared. That is the sort of outcome that makes a pathway feel real. It does not rely on the animal's choices in a behavioural arena. It says: when the magnetic environment changes, neurons in a particular sensory circuit are measurably more active, and that activation depends on a specific nerve.

Comparable ZENK mapping has also been reported in pigeons under changing magnetic fields, again with activation patterns in trigeminal brainstem nuclei that vanish under conditions designed to remove magnetic stimulation or after transection of V1. These studies matter because they anchor a controversial behavioural literature to a physiological substrate. Even if there is debate about the exact receptor cell in the beak, the trigeminal pathway behaves as though it is receiving some kind of magnetically modulated input.

At this point, a reasonable person might ask a disarming question: why should a "map" be carried by the trigeminal nerve at all? Why not by the same system that might underlie a directional compass? The answer is partly anatomical and partly computational.

Anatomically, a directional compass in birds is often discussed as a light-dependent process associated with the eye and visual system, whereas the trigeminal system is classically a somatosensory route. If there are two magnetic senses—one for compass direction and one for map-like information—it is plausible that evolution would build them on different scaffolds, especially if the underlying physics differs. A compass based on light-dependent chemistry naturally belongs near photoreceptors. An intensity-sensitive map based on magnetite-based mechanotransduction can be placed in a peripheral tissue, near mechanosensitive endings, and sent inward through a nerve specialized for facial sensation.

Computationally, intensity information is not a direction; it is a scalar. The Earth's field strength changes gradually across geography, and so does the inclination angle, the tilt of the field lines relative to the surface. An animal can potentially use these gradients, along with other cues, to estimate its position relative to a goal. A scalar signal can be carried by many receptors without demanding a strict anatomical alignment like a compass needle does. It can also be integrated with olfactory cues, wind cues, and learned landmarks. A trigeminal channel, with its dense peripheral innervation and robust brainstem processing, could plausibly support such integration.

There is, however, a discomfort at the heart of the trigeminal map story: the receptor itself has been hard to pin down. For years, one prominent candidate was a set of iron-rich structures in the upper beak region of pigeons, described as if they were arranged in a way suitable for magnetic transduction. The picture was seductive: iron minerals placed close to nerve endings, neatly positioned for the Earth's field to tug upon.

Then the ground shifted. Christoph Treiber, Marion Salzer and colleagues published a careful study in 2012 arguing that the conspicuous iron-rich cells in the pigeon upper beak are not specialized magnetoreceptors at all, but macrophages—immune cells that ingest debris and can accumulate iron. Macrophages, after all, are meant to be variable in number and location because they respond to local tissue history. A stereotyped sensory organ should not wander about the way immune cells do. The implication was not merely that one group had made a mistake, but that the most visually satisfying "magnetic organ" in the beak might be an illusion created by the fact that iron is biologically ubiquitous and histologically treacherous.

Henrik Mouritsen's commentary around the same time captured the tone of the moment: the search for compass needles in tissue had to be held to the same standards as any sensory receptor. There must be a plausible link to neurons. The structure must be reproducible. It must be demonstrably connected to sensation rather than to housekeeping.

This is the kind of scientific upset that is both chastening and productive. It does not eliminate the trigeminal pathway evidence; if anything, it sharpens it. If cutting V1 eliminates magnetic-field-dependent activation in the brainstem, something in the periphery is sending magnetic information through that nerve. If the obvious iron clusters are macrophages, then either the true receptor is rarer, subtler, or located elsewhere, or the signal arises through a mechanism that does not leave an easy mineral fingerprint. The burden of proof shifts from "here is a big iron stain" to "here is a specific transducing unit with neural identity".

The debate also forces a cleaner separation between two different claims that are sometimes conflated. One claim is *pathway*: the trigeminal system carries magnetically modulated signals. Another claim is *substrate*: the peripheral receptor contains magnetite (or another ferromagnetic mineral) and transduces magnetic forces mechanically. The pathway claim is supported by multiple lines of behavioural and physiological work. The substrate claim is physically plausible and supported in other organisms (magnetotactic bacteria are the obvious example), but in birds the anatomical evidence has been more contested.

This distinction matters for another reason: nerve signals have a "language", and by paying attention to that language one can infer what the receptor might be encoding. Electrophysiological recordings provide a direct view. In the 1990s, recordings from the ophthalmic nerve and trigeminal ganglion in a migratory songbird reported units whose firing rates changed in response to small variations in magnetic intensity, in some cases down to changes of a few hundred nanotesla. That is a tiny perturbation relative to the Earth's field, which is tens of microtesla. A sensitivity of that order is not absurd for a biological sensor, but it is demanding. It implies either an efficient transducer or an effective neural strategy for extracting signal from noise.

If the trigeminal pathway conveys intensity information, the practical question becomes: what does "intensity" feel like to a brainstem nucleus? Not as a number printed on an instrument, but as a shift in spike timing or firing rate. A receptor cell could increase its receptor potential as intensity increases, leading to more spikes in its afferent fibre. Or it might be tuned to changes, responding strongly when intensity increases or decreases, much as some touch receptors respond to movement rather than steady pressure. The ZENK studies that use *changing* magnetic fields rather than steady fields hint that dynamics matter: the system may be especially responsive when the magnetic environment varies, as it naturally does when a bird moves through space.

This offers a satisfying picture of how a "map" could begin. A bird flying over unfamiliar terrain experiences a changing pattern of magnetic intensity and inclination as it moves. Those changes are not unique coordinates, but they provide constraints. Coupled with wind direction, time of day, odours, and learned expectations, magnetic gradients can help an animal decide whether it is getting warmer or colder relative to home. The trigeminal pathway, with its strong connections to brainstem networks involved in sensory integration and orienting behaviour, is well placed to contribute such constraints.

A concrete example makes the logic less abstract. Imagine a homing pigeon released in a region where the magnetic intensity is slightly higher than at its loft. If the pigeon has learned, through experience, that intensity decreases on the way home along typical routes, then an intensity-sensitive trigeminal channel could bias its initial search. Now imagine that the same pigeon is released in a location with similar intensity but a different pattern of landmarks and smells. Magnetic intensity alone would not tell it the answer, but it might reduce the space of plausible directions, making the bird's early decisions less random. The "map" is not a full coordinate system; it is a set of cues that can be weighted differently depending on reliability.

This is also why nerve sectioning experiments have been so influential in separating compass from map. When the trigeminal ophthalmic branch is cut, birds can still show directional orientation under some conditions, suggesting that the compass-like component can remain intact. But their ability to respond to magnetic anomalies or to make use of magnetic intensity cues can be degraded. This double dissociation is never perfect—biology rarely offers clean separations—but it is persuasive. It suggests that at least two magnetic information channels exist, and that they enter the brain through different doors.

The trigeminal story also reconnects magnetoreception to a broader theme in sensory biology: many senses are composites. Even human taste is not a single modality; it is taste plus smell plus texture plus temperature. Animal navigation is similar. It is not a "magnetic sense" competing with a "sun compass", but a nervous system that arbitrates among cues that can be reliable one day and misleading the next. A trigeminal magnetic channel becomes particularly valuable in conditions where visual cues are degraded: overcast skies, featureless landscapes, nighttime travel, or long ocean crossings. It can also act as an internal consistency check. If the compass points one way but the intensity trend suggests another, the animal can avoid committing too strongly to a single cue.

For readers who have never thought about their own trigeminal nerve, there is a revealing exercise. On a cold day, inhale sharply through the nose. The sting you feel is not "smell" in the usual sense; it is trigeminal. Menthol, ammonia, the burn of chili—these are trigeminal sensations, not purely olfactory ones. The nerve is tuned to chemical irritation, temperature, and tactile flow. It is built to detect the physical properties of the world as they touch the face. In that context, the idea that it might also carry information about an invisible physical gradient is less odd than it first appears.

The remaining mystery is where, exactly, the first transduction occurs. If magnetite-based receptors exist in the beak region, they must be associated with neural elements in a way that survives the macrophage critique. They may be sparse, distributed, embedded in connective tissue, or located in structures not easily revealed by traditional stains. Or the receptor may use iron minerals in a form that does not stand out as dramatic crystals. The search has become more careful and, in a sense, more interesting: the field is no longer content with "iron is present", but demands "iron is wired".

That demand is not pedantic; it reflects the difference between material and meaning. A magnetite particle in a macrophage is part of the immune system's housekeeping. A magnetite particle coupled to a mechanosensitive ending and connected through V1 to PrV is part of a sensory experience. The trigeminal map, as a hypothesis, lives in that wiring diagram. It is the story of how a cue from the planet becomes a pattern of spikes, and how a pattern of spikes can help a bird decide which way to turn its head when the horizon offers no obvious answer.

When a pigeon lifts off after release, it often circles, climbs, and then sets a bearing that looks purposeful even to a human observer standing in a field. That purposefulness is not magic. It is the nervous system doing its quiet arithmetic: weighing wind on the feathers, the smell of familiar countryside, the brightness of the sky, and perhaps a faint, persistent signal arriving from the beak along a nerve best known for pain and touch. In the brainstem, among circuits that evolved to keep food out of the nostrils and to warn of cold air, magnetic information can be treated like any other bodily sense: something that can be trusted, doubted, combined, and, when the world demands it, acted upon.

Chapter 5

# The Great Aerial Odyssey

*Every autumn, billions of songbirds embark on journeys that span hemispheres, guided by an invisible force that has baffled humans for centuries. By bringing the European robin into the laboratory, scientists have begun to map the neural circuitry that transforms magnetic fields into navigational commands. This chapter unravels the specific experiments that proved birds do not just sense the north-they likely see it.*

## 5.1 The Robin in the Funnel

A caged bird in spring can look perfectly calm. Put the same bird in a quiet room on an October night and it may turn into something else entirely: a small engine of impatience, hopping and fluttering as if the walls were a mistake that should be corrected by force. German ornithologists gave this migratory agitation a name that sounds like it belongs in a fairy tale: *Zugunruhe*, literally "migration restlessness." It is not random panic. It has a timing, a rhythm, and—most intriguingly—a direction.

For centuries, people inferred migration from absences and arrivals, from nets that filled and then went quiet, from the uncanny way a place could be full of song one week and empty the next. The question of *how* a bird keeps a course across a featureless sea, or through cloud, or under a blank canopy of stars, demanded a different kind of evidence: not the grand gesture of a journey completed, but the small, repeatable gesture of a decision made. A migrant, even one confined to a laboratory, carries a program that can be coaxed into showing its hand.

The cleverness of the Emlen funnel is that it asks the bird a question in the only language the bird can answer with certainty: movement. It does not require the animal to peck a key, or learn a rule, or understand a reward. It does not pretend to recreate a thousand-kilometre flight. It simply gives a restless bird a stage, then records the direction in which that restlessness strains.

Stephen T. Emlen and John T. Emlen introduced the method in 1966, at a time when the very idea of a laboratory assay for migration could have sounded like trying to bottle weather. Their device is almost disarmingly simple: an inverted cone of paper, a perch, a view of the sky (or an artificial sky), and the animal's own feet as the pen that writes the result. In its classic form, a small songbird is placed inside the funnel at night during migratory season. The bird tries to take off—not with sustained flight, which a cage prevents, but with repeated hops up the sloping wall. Each hop is a vote. If the votes cluster, the bird has expressed a preferred direction.

What makes the funnel more than a behavioural curiosity is its ability to turn those votes into data that can be counted, compared, and challenged by controlled manipulations. Direction, in science, has to be measurable. "Northwest" cannot remain a poetic claim; it has to become an angle on a circle, with an uncertainty you can argue about.

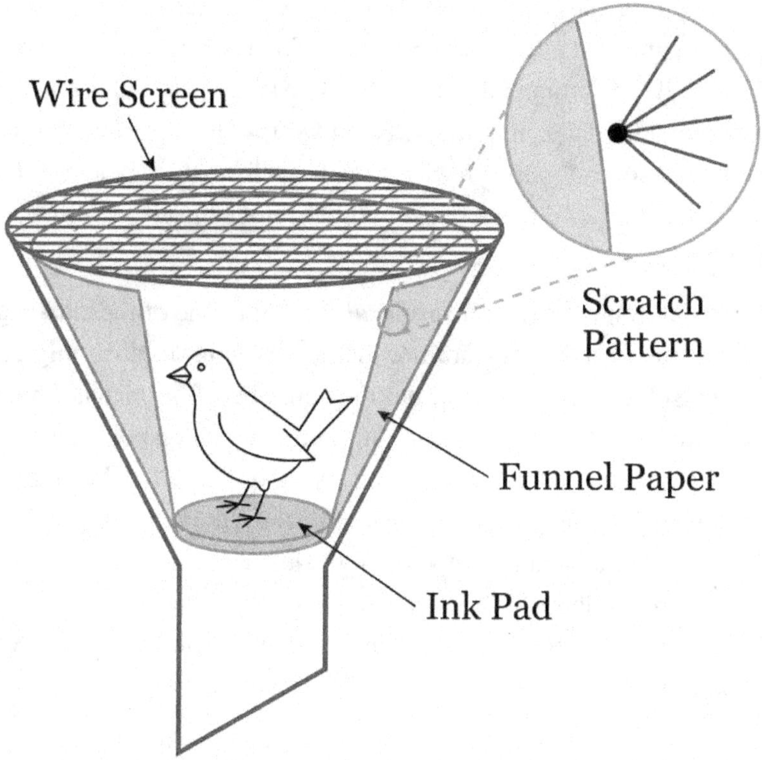


The funnel works because the bird's attempt to depart is not a single leap but a repeated pattern. A migratory songbird in a funnel typically faces up the slope, hops, lands, hops again. Each hop carries a directional intention: the bird is trying to launch toward a particular sector of the horizon. The paper lining the inside wall records this, often with the help of an ink pad placed on the floor or a light coating of pigment on the bird's feet. When the bird hops, it leaves short marks where its toes or claws skid and scrape. In the morning, the experimenter removes the paper and counts the marks around the circle.

The image can be surprisingly human in its legibility. A well-oriented bird does not scribble everywhere; it concentrates its efforts. A spring migrant might produce a dense fan of scratches aimed northeast, while an autumn migrant of the same species, tested in the same room months later, might lean southwest. This seasonal reversal is one reason migratory restlessness is such a gift to science: it is not merely arousal but a structured behaviour linked to an internal program.

European robins (*Erithacus rubecula*) became one of the classic actors in this story. They are not glamorous long-distance migrants on the scale of Arctic terns, but they are abundant, manageable in captivity, and strongly seasonal in their orientation. Anyone who has watched a robin in a garden may think of it as a local resident, a bird of hedges and lawns. In much of Europe, though, many individuals do migrate, and their nocturnal restlessness can be elicited with remarkable reliability. That reliability matters, because a compass sense cannot be studied in a creature that refuses to show you where it wants to go.

There is a small but telling historical anecdote that captures the tone of early funnel work. Researchers would sometimes run trials in buildings not designed for delicate sensory experiments: rooms with unseen metal pipes, electrical noise, and the faint vibrations of a campus at night. A batch of birds might suddenly "lose" its direction, scattering scratches everywhere. The easy explanation would have been to blame the animals: perhaps they were stressed, perhaps they were unmotivated. But the Emlen funnel makes it possible to suspect the room instead. If the same birds recover their neat directional pattern after a change in lighting, or after a shift in the magnetic field produced by coils, the experiment has taught a deeper lesson: orientation is not a mood; it is information processing.

The funnel by itself is a recording device, not a compass. The compass question enters when the animal is tested under altered cues.

The most dramatic cue to manipulate is the sky. Emlen's original work was closely tied to celestial navigation: birds placed under a planetarium sky, where the star pattern could be rotated, would rotate their preferred direction accordingly. The funnel made that result crisp. The bird was not merely becoming more active; it was changing its heading in a lawful way.

The same logic later proved crucial for studying the geomagnetic field, which has the awkward property of being everywhere and usually invisible. To test whether a bird uses magnetism, one wants to change the magnetic field without changing anything else. In practice, this is done by placing the funnel inside a set of coils that can add a controlled magnetic field to the Earth's field, tipping it, rotating it, or altering its inclination (the angle at which field lines enter the ground). A bird that follows those manipulations is doing something that deserves to be called magnetic orientation.

Here the funnel's simplicity becomes a strength. Because the bird cannot fly and cannot see landmarks, and because the apparatus can be placed in a uniform, featureless enclosure, the experimental world can be stripped down to a few cues at a time. When the bird nonetheless produces a consistent direction, the direction has to come from something it can sense inside that stripped-down world.

The geometry of the Earth's magnetic field is worth holding in mind. Field lines emerge from the ground in one hemisphere, arch through space, and re-enter in the other. Near the magnetic equator, the field is close to horizontal; near the poles, it is steeply inclined. A compass needle aligns with the field's horizontal component and points toward magnetic north or south depending on its polarity. Human compasses, as every scout is told, are polarity compasses: they distinguish north from south by the "north-seeking" end of the needle.

Robins do not work that way. The funnel helped make this clear through a series of experiments that now read like a detective story, with the Earth's field as the suspect and the bird's feet as the witness.

Wolfgang and Roswitha Wiltschko's work in 1972 with European robins is central here. Using controlled magnetic fields and behavioural assays like the funnel, they showed that the birds orient using an *inclination compass.* Instead of detecting polarity (north versus south), an inclination compass detects the *axis* of the magnetic field lines and whether they tilt downward or upward relative to gravity. In the Northern Hemisphere, field lines angle downward toward the north; in the Southern Hemisphere, they angle downward toward the south. A polarity compass can tell you "toward the magnetic North Pole" as a unique direction. An inclination compass is subtler: it can tell you "poleward" versus "equatorward" along the field axis, but it does not care which end of the axis is labelled north.

Why would evolution settle for this apparently poorer instrument? The answer is that it is not poorer for a migrant; it is robust. An axial compass based on inclination is naturally compatible with a sensory mechanism that responds to the *alignment* of an internal process with the field, rather than to the field's sign. Many chemical and physical processes in tissues respond symmetrically to a field reversal: flip the field, and the process behaves the same. If the bird's sensor is built on such a process, the animal gets an axial compass almost for free. It also gets a compass that works in both hemispheres without requiring an internal "switch" for polarity.

The funnel makes the inclination compass concept concrete. Imagine a robin that, in autumn, scratches mostly toward the southwest when tested in the local geomagnetic field. Now reverse the horizontal component of the magnetic field with coils, effectively swapping magnetic north and south while leaving the inclination the same. A polarity compass would reverse its heading; a pure inclination compass should not. The birds behave as if they are tracking the inclination-defined axis rather than polarity. More decisive still are manipulations that invert the vertical component, changing the inclination from downward to upward. When the inclination is inverted, birds shift their orientation in a way consistent with "poleward" versus "equatorward" being defined by the sign of the inclination relative to gravity, not by polarity. Without a device like the funnel, such claims would remain airy. With it, they become scratches on paper, aggregated across nights and individuals until a pattern emerges from the noise.

It is easy to underestimate how psychologically strange this is until one tries to imagine it from the inside. A human can feel "up" and "down" through the vestibular system in the inner ear, and can feel gravity through muscle and joint sensors, but cannot feel the direction of magnetic field lines at all. A robin, by contrast, appears to combine an internal sense of gravity with a magnetic cue that is not a simple arrow but an angle—the slope of an invisible structure threading the world. The bird's "north" is not the same concept as ours. It is closer to a rule: move in the direction where the field dives, or away from it, depending on season.

The funnel does more than reveal the compass type; it also forces a discipline on interpretation. Migratory orientation is noisy. A bird might be active but scattered; another might sit still; a third might show two clusters, as if torn between alternatives. The Emlens' contribution was not merely an apparatus but an experimental grammar: test many individuals, quantify directional distributions, and accept that behaviour must be treated statistically. One robin's scribble is a story; thirty robins' scribbles can be evidence.

The method also encourages humility about what a "decision" means. In the wild, a migrant does not choose a heading once and keep it forever. It adjusts to wind, terrain, weather, fatigue. In a funnel, none of that is possible. What is measured is a *preferred bearing* under a particular set of cues, at a particular time of night, in a particular physiological state. That limitation is real, but it is also precisely what makes the funnel powerful: it isolates the compass component of navigation from the messy business of travelling.

To appreciate the device's elegance, consider the alternative. Suppose one tried to test magnetic orientation by releasing birds in a field and watching where they flew. Some would vanish behind trees. Some would circle. Some would land. Weather would change. Landmarks would intrude. The Earth's magnetic field could not be rotated for half a hectare without absurd engineering. The funnel compresses a night sky into a manageable aperture, and compresses a migratory urge into a set of marks you can hold in your hands the next morning.

This is why the robin in the funnel became a sort of quiet protagonist in magnetoreception research. When later ideas appeared—that the magnetic compass might depend on light, that certain wavelengths might be necessary, that radio-frequency noise might disrupt the sense—the funnel was ready. A small change in the lighting over the funnel, or a small antenna emitting a controlled field, could be paired with the same behavioural readout. If the bird's directional scratches dissolve into confusion under a particular condition, and then recover when the condition is removed, the result has a satisfying clarity.

The funnel can also be used to stage cue conflicts, which are among the most revealing experiments in animal navigation. Give a bird both a star pattern and a magnetic field, then rotate one relative to the other. Which cue wins? Does the bird compromise? Does it switch depending on season, experience, or time of night? Because the funnel produces an orientation distribution, not merely a yes-or-no response, it can reveal compromise headings and bimodal choices. It lets the bird show uncertainty, and uncertainty is often where the mechanism is hiding.

None of this should obscure the almost domestic intimacy of the setup. In many laboratories, the evening routine has an almost ritual character. The room is prepared: extraneous lights taped over, equipment checked, coils calibrated. A robin is carried in gently, placed inside the funnel, and allowed to settle. The bird is not being trained; it is being given a chance to express something it already knows how to do. For hours it may hop and flutter, a small heart beating fast in the dark, its muscles enacting a journey it cannot take.

In the morning, when the paper is unrolled, the night's invisible computations become visible. A fan of scratches points somewhere beyond the walls. It might be neat and narrow, or broad, or split. It might shift when the field is altered, and shift back when the natural field is restored. On good nights, the result feels less like a trick of confinement than a message: even without a horizon, even without a landscape, something inside the bird can take a global cue and turn it into direction.

That message is both sobering and oddly reassuring. Sobering, because it reminds us that the senses we think of as complete are merely local. Reassuring, because it suggests that the world is richer in structure than we notice. The Earth's magnetic field is not just a property of compasses and iron filings; it is part of the lived environment of other creatures, as real to them as the line of the horizon is to us. The Emlen funnel, made of paper and mesh and patience, is a way of borrowing that perspective for a moment, long enough to see direction appear where our own senses would swear there is only dark.

## 5.2 The Neural North

A funnel can tell you *that* a robin chooses a direction. It cannot tell you *where* in the animal that choice is computed. For a long time, the magnetic sense had the feel of a rumour: persuasive behaviour on one side, plausible physical mechanisms on the other, and a gap in the middle where nerves and synapses ought to be. The most unsettling possibility was that the gap was permanent, that magnetoreception might turn out to be one of those real biological abilities that resists localisation—spread diffusely through the body, or expressed only as a whole-organism effect. Then a very specific patch of forebrain began to light up at exactly the right moment.

The story begins with a methodological trick that neuroscientists sometimes use when they do not know what to look for. Rather than guessing the location of a "magnetic organ" and poking around, they let the animal's own brain reveal what it is doing. Neurons that are strongly active can switch on certain genes within minutes, producing messenger RNA and proteins that can be stained in thin sections like footprints after the fact. One of the commonly used markers is a gene called *ZENK* (also known under other names in different species), which becomes expressed in many kinds of neurons after activation. The idea is not that *ZENK* is "the navigation gene"; it is simply a convenient lantern. Wherever the lanterns glow, something important happened.

In 2005, working with night-migratory songbirds, researchers mapped this activity and found a surprise: a small region in the forebrain that was quiet during the day and in non-migratory periods, but became highly active at night specifically when the birds were in migratory mode. It was not a broad wash of activity across the visual cortex, or a general arousal signal; it was a cluster. The name it acquired, "Cluster N," is almost comically plain for something so evocative: N for "night." It sounded like an internal landmark, a place where nocturnal migrants might be doing a special kind of seeing.

The first temptation was to treat Cluster N as the long-sought compass centre. But brains rarely offer such gifts without strings attached. A region that lights up at night might be responding to dim light rather than to magnetism. It might be involved in the general restlessness of migration, or in sleep suppression, or in heightened attention. To make the case that Cluster N had anything to do with the magnetic sense, one needed two kinds of evidence: a plausible pathway from the eyes, where many suspected the compass began, and a functional test showing that disrupting the cluster specifically disrupts magnetic orientation.

Both arrived in the following years.

Anatomy, in neuroscience, is not merely a diagram of parts. It is a set of constraints. If information from the eye is going to influence a forebrain region, there has to be a route—a series of neurons connected like relay runners. Birds have more than one major visual pathway. Humans learn the textbook route: retina to lateral geniculate nucleus to primary visual cortex. Birds share some analogous structures, but their wiring is not a simple copy. One prominent route in birds is the *thalamofugal* pathway, which carries information from the retina to a thalamic nucleus and then to a dorsal forebrain region often called the visual Wulst. Another route, the *tectofugal* pathway, runs through the optic tectum and is crucial for many kinds of visually guided behaviour. If the magnetic compass is "in the eye" in the sense of beginning with photochemistry in retinal cells, then any central processing would likely ride along one of these visual highways.

Dominik Heyers and colleagues (2007) traced connections consistent with this idea. Cluster N sits in the visual Wulst and receives input through the thalamofugal pathway. In other words, the region that becomes active during nocturnal migration is not an isolated nocturnal island; it is embedded in a visual network, positioned exactly where a specialised visual computation could be carried out. Even more suggestive, Cluster N activation depends on visual input: cover a bird's eyes, and the activity collapses. That result alone does not prove magnetoreception, but it narrows the possibilities. Whatever Cluster N is doing, it is doing it with the eyes open.

That might sound obvious until one remembers the older, competing intuition about magnetoreception: perhaps there is iron in the beak, perhaps there are magnetic particles acting like tiny compass needles, perhaps the trigeminal nerve that innervates the face is carrying magnetic information to the brainstem. Those ideas have their own evidence and remain relevant, but Cluster N makes it harder to imagine the magnetic compass as a purely beak-based sensor. A forebrain region embedded in the visual system, active only when migrants are orienting at night, begs for an explanation that involves light.

The most influential physical hypothesis along those lines is the radical-pair mechanism. The phrase sounds like chemistry class, and it is, but with a twist of physics. Certain molecules, when struck by light, can enter a state where they have two unpaired electrons whose spins are correlated—a "radical pair." The fate of that pair, whether it recombines or proceeds to other chemical products, can depend on the orientation of the molecule relative to an external magnetic field. In the retina, proteins called cryptochromes have been proposed as candidates for generating such radical pairs. The important conceptual point is that this kind of sensor is not a compass needle. It does not point. It modulates a chemical reaction in a way that depends on angle. If the output of that chemistry influences retinal signalling, then the bird might experience magnetism as a subtle pattern superimposed on vision: an *overlay* rather than an arrow.

Cluster N fits neatly into this picture. A magnetic overlay would need interpretation. It would need circuitry that can compare signals across the retina and turn a field-dependent modulation into a stable directional reference. It would need, in effect, a visual brain region that treats magnetic information as a special kind of visual cue.

That is still an inference, and the brain is good at punishing inferences that go untested. The decisive question was behavioural: if Cluster N is disrupted, does the bird lose its magnetic compass while keeping other forms of orientation intact?

Zapka, Heyers, and colleagues (2009) addressed this in European robins with a kind of experiment that feels almost surgical in its logic. The same animal can be tested under different compass systems. A robin can orient using the magnetic field under appropriate conditions, but it can also orient using stars, and it can use the sun when available. If a brain region is part of a general "navigation centre," damaging it should impair all these. If a region is specifically tied to magnetic compass processing, then magnetic orientation should fail while celestial orientation remains.

They inactivated Cluster N and found that magnetic compass orientation was abolished—the birds no longer showed the orderly directional preference one expects—yet the same birds could still orient using stars or the sun. This is the kind of dissociation that neuroscientists treasure because it is difficult to explain away. The birds were not simply confused, not simply lethargic, not globally impaired. They had lost one particular ability while keeping others. It is also the kind of result that makes a popular account tempting to overstate: "the magnetic centre discovered!" But even here, restraint is warranted. Cluster N may not be the *source* of magnetic sensing; it may be a crucial processing station, a place where magnetic information, generated elsewhere, is rendered usable for behaviour.

The experiment does something else, quietly but powerfully. It tells us that a "magnetic compass" is not merely a property of molecules; it is a property of an entire system, from receptor to brain to motor output. When the system is intact, a robin in a funnel produces a coherent direction. When a specific neural node is knocked out, the same animal can still see, still move, still express restlessness, yet cannot make magnetically guided choices. The compass has become a neural function rather than a mystical faculty.

The link to the eye becomes still more intriguing when one considers the timing. Cluster N becomes highly active during nocturnal migration, when light levels are low. A magnetic overlay, if it exists, must operate in dim conditions, and it may impose constraints. Indeed, behavioural work across species suggests that magnetic compass orientation in birds depends on light and can be disrupted under certain wavelengths or intensities. That observation is often taken as a sign of a light-dependent, cryptochrome-based mechanism. But the nervous system has to cope with the fact that nocturnal light is not a laboratory lamp: it varies with moon phase, cloud cover, canopy, and artificial lighting. A special visual region active during migration could be a way to adjust gain, filter noise, and extract a weak, field-dependent signal from a fluctuating visual environment.

Here an everyday analogy helps, though it is imperfect. Imagine trying to read a faint watermark in paper. In bright light, the watermark vanishes; the glare overwhelms it. In dim light, it is easier to see, but only if your eyes and brain adjust. You might tilt the page, change your angle, squint, ignore irrelevant texture. The watermark is always there, but perceiving it is an active process. If magnetoreception is a kind of watermark on vision, Cluster N may be part of the brain's effort to make that watermark legible at night.

A concrete example makes the stakes clearer. Consider a young robin on its first autumn migration. It has never been to Spain or Morocco. It has no map in the human sense, no memory of a route. Yet its migratory program pushes it in a consistent direction. If the magnetic compass is part of that program, then an internal visual overlay would have to be interpretable without learning a landscape. The direction should emerge from an innate computation: align the body until the overlay matches a desired pattern, then fly. A specialised nocturnal visual circuit could be the neural embodiment of that innate computation.

Cluster N also forces a new kind of question about what "seeing" means. In human vision, early visual cortex contains maps: neighbouring neurons respond to neighbouring points in space. Features such as edges and orientations are extracted in a systematic way. If the magnetic field is represented as a visual modulation, it might create patterns across the retina that depend on head orientation relative to the field lines. The brain could then treat those patterns as it treats other spatial patterns, building a sense of direction from a stable relationship between the pattern and the body.

This line of thought makes a prediction: change the bird's visual input and you should alter the magnetic compass, not because the eyes provide landmarks, but because the compass itself is light-dependent. In fact, in the 2005 work that first identified Cluster N, eye-covering reduced the activation dramatically, suggesting that the region is driven by visual stimulation rather than by an internal migratory clock alone. The behavioural dissociation work (2009) then made the argument functional: it is not merely that Cluster N is a night-vision area; it is needed specifically for magnetic compass orientation.

At this point, a skeptical listener might ask a reasonable question at a dinner party: if Cluster N is visual and magnetic, why do birds not constantly use it, day and night? Why make a special region that flares only during migration? Part of the answer may be that the magnetic compass is not always needed. A bird foraging in familiar territory can use landmarks. The cost of maintaining a high-gain, noise-sensitive sensory mode might outweigh the benefits outside migration. Another part of the answer may be that the magnetic overlay, if it depends on cryptochrome photochemistry, is suited to particular light conditions and may compete with ordinary visual tasks. A system tuned to extract a weak, field-dependent signal might not be optimal for high-acuity colour vision in daylight. The brain can, in effect, choose its sensory priorities.

The trigeminal pathway, often associated with beak innervation, complicates the picture without negating it. Other studies have reported that changes in magnetic fields can activate trigeminal brainstem complexes in migratory birds, and that the ophthalmic branch of the trigeminal nerve seems involved in some magnetic responses. This suggests that birds may have more than one magnetic channel: one that supports a compass (direction) and another that could contribute to a map (position), perhaps by sensing intensity or gradients. The funnel experiments that demonstrate an inclination compass speak to direction; Cluster N speaks to processing that direction in the visual forebrain. The trigeminal evidence suggests that magnetism may also be monitored as a bodily signal, not only as a visual overlay. A navigational animal would hardly be unusual in having redundant sensors. Airplanes do. So do people, who can infer orientation from vision, vestibular sensation, proprioception, and even sound.

The European robin, though, has become a focal point because its behaviour is cooperative and its neurobiology is tractable. One can run the same basic behavioural assay night after night, then pair it with anatomical tracing, gene-expression mapping, and reversible inactivation. A set of questions that once belonged to natural history and speculation can be brought into a laboratory routine.

There is also a human story in the background, one that rarely appears in papers but shapes what gets discovered: the patience of working at night with small animals that did not consent to be your model organism. The discovery of Cluster N depended on sampling brains at the right times, under the right seasonal states, with careful control of light exposure. It is not hard to imagine the rhythms of that work: late hours, dim rooms, birds that migrate on their own schedule rather than yours. The word "cluster" understates the labour that turned faint molecular staining into a coherent map.

In the end, Cluster N offers a kind of neural north not because it is a needle pointing to the pole, but because it is a signpost that the magnetic sense is handled by the brain as a specialised sensory modality. Behavioural scratches in a funnel become, through the bridge of neuroanatomy, a pattern of activity in a particular part of forebrain, tied to a visual pathway, switched on at the moment when the animal most needs an internal reference. When a robin is poised to leave familiar places behind, the night sky is not the only expanse it consults; there is also an invisible geometry passing through its head, rendered into neural signals that converge in a small patch of tissue doing the quiet work of making direction feel like a fact.

## 5.3 The Solar Calibration

A compass, on its own, is an optimist. It assumes that whatever it points to is worth following, and that the creature holding it can interpret it correctly. In the lab, a robin can be asked to choose a direction under a carefully arranged magnetic field, and it will oblige with scratches on paper. In the living world, the same robin must solve a harder problem: it must keep its internal compasses honest.

Honesty is not a moral virtue in navigation; it is a technical one. Every cue an animal uses can drift. The sun moves across the sky. Stars rotate. The magnetic field varies with location and can be locally distorted by geology or human infrastructure. Even the animal itself changes: hormones rise and fall with season, sleep is interrupted, attention is pulled by hunger or predators. A navigation system that relies on one cue with unshakable faith would be brittle. Birds are not brittle. They are, instead, calibrators.

Calibration is what you do when you have several instruments, none perfect, and you let them check each other. A pilot trusts the compass less when flying near metal; a sailor checks a compass against the sun. A bird does something analogous, but in a more intimate way, because its instruments are biological. Somewhere between the retina and the forebrain, between muscle and inner ear, the animal is constantly aligning its internal references. The magnetic compass is one reference. The sky is another. The circadian clock—the internal timekeeper that tells the bird whether it is morning, noon, or evening—binds them together.

The key insight behind decades of work on avian orientation is that birds do not treat the magnetic field as a dictator. They treat it as a partner in a coalition of cues, and the coalition has a pecking order that can change with context. At night, when many songbirds migrate, the magnetic compass is particularly useful: it is available under clouds and over dark sea, and it does not care about light pollution in the same way that stars do. Yet even a magnetic compass can be misled if it is not anchored to the wider world. A compass that has drifted by fifteen degrees is not broken; it is simply dangerous.

One of the most revealing ways to watch calibration happen is to make the cues disagree on purpose and see what the bird does. These are cue-conflict experiments, and they are the behavioural equivalent of optical illusions: carefully constructed situations where a reliable sensory system is invited to reveal its assumptions. The Emlen funnel makes such experiments possible because it supplies a clean readout of directional preference. The coils that manipulate magnetic fields and the devices that manipulate celestial cues supply the controlled conflict.

A classic conflict is between the magnetic field and the star pattern.

Place a bird under an artificial sky in a planetarium and rotate the star pattern while leaving the magnetic field unchanged, or rotate the magnetic field while leaving the stars unchanged. A bird that follows the stars is revealing that its star compass is currently dominant; a bird that follows the magnetic field is revealing magnetic dominance. More interesting still is what happens after the conflict is removed. If a bird is exposed to a rotated star pattern at dusk and then tested later under a normal sky, it may have shifted its magnetic orientation as if it has *recalibrated* the magnetic compass to match the stars. That is calibration in action: not simply choosing one cue over another in the moment, but updating the internal relationship between them.

The time of day matters because the sky contains a special piece of information at twilight. Humans notice twilight for its colours, but birds notice it for its geometry. Around sunrise and sunset, the pattern of polarized light in the atmosphere becomes especially structured. Polarized light is light whose waves are aligned more in one plane than another, a property that many animals can detect even though we mostly cannot without special filters. In the sky, polarization patterns arise from scattering of sunlight by molecules; the pattern forms arcs and bands that relate to the position of the sun, including when the sun is just below the horizon. For a bird, twilight is not merely dimming or brightening. It is a brief, dependable calibration display that the atmosphere puts on twice a day.

This matters because a sun compass is not a simple compass. The sun moves. If an animal uses the sun's position to infer direction, it must compensate for that movement, or it will fly a curve. The compensation comes from the circadian clock. A time-compensated sun compass works like this: if the bird knows the time of day, and it knows how the sun's azimuth changes with time, it can translate the sun's current position into a stable direction such as geographic north. Without an accurate internal clock, the sun compass becomes systematically wrong.

The elegance of the calibration idea is that it ties these pieces together. The bird can use the sky at twilight to calibrate the magnetic compass, and it can use its circadian clock to interpret the sun at other times. In effect, the bird builds a stable frame by forcing agreement among magnetic cues, celestial cues, and internal time.

The evidence that birds really do rely on their clock when using the sun comes from a beautifully simple manipulation: shift the clock. In clock-shift experiments, a bird is kept on an artificial light-dark cycle that is advanced or delayed relative to local time. The bird's internal clock adjusts. Then the bird is released or tested under the real sun. Because the bird's clock is wrong relative to the sun's actual position, its sun compass produces a predictable error. If the clock is shifted by six hours, the sun compass heading shifts accordingly. This is not mere disorientation; it is a lawful mistake, the kind of mistake a system makes when it is operating correctly with incorrect input.

Such experiments, developed and refined in the mid-twentieth century, have a human flavour that is hard to miss. The bird is, in effect, given jet lag without travelling. The predictable nature of its error is what reveals the underlying computation. When a bird aims too far to one side by an amount that matches the clock shift, it is as if the animal is saying: "I am using the sun with the time you gave me."

Now place the magnetic compass into this web. The magnetic compass does not, in principle, require time compensation; the magnetic field does not rotate through the day like the sun does. But the magnetic compass can still benefit from time because the calibration opportunities are time-locked. Twilight comes at predictable times. The bird's migratory programme also has predictable timing: nocturnal migrants become restless at night, not at noon. The circadian clock is therefore not merely a sun-compass accessory; it is a scheduler that determines when the brain enters certain sensory modes and when certain kinds of learning or recalibration may occur.

A concrete example helps to make calibration less abstract. Imagine an autumn night when a young European robin is preparing to leave. It has spent the day foraging in hedgerows and gardens, using ordinary vision and memory. Near sunset, the sky offers polarized patterns and the sun's position relative to the horizon. The bird can compare what its magnetic compass says with what the twilight sky implies about direction. It does not need to understand astronomy. It needs only to link two sensory states: a particular internal magnetic "north" with a particular twilight configuration. If the magnetic compass has drifted—perhaps because of local magnetic anomalies, perhaps because the bird has been near a metal structure, perhaps because its neural processing has been noisy—twilight provides a chance to nudge it back into alignment.

Laboratory work suggests that such recalibration can be surprisingly quick. Under controlled conditions, birds exposed to altered magnetic fields during twilight can later show shifted magnetic orientation, as if they have accepted the twilight cue as the reference and adjusted the magnetic compass accordingly. Conversely, if the relevant twilight cues are blocked or altered, the recalibration may not happen. The implication is not that birds constantly re-learn north; rather, they have a built-in habit of checking.

This habit also helps resolve a subtle tension in magnetoreception. The avian magnetic compass, as inferred from inclination-compass behaviour, is not a polarity compass. It distinguishes poleward from equatorward based on the tilt of field lines relative to gravity. That tells a migrant which way is toward higher latitudes, but it does not automatically deliver geographic north in the way a human imagines it. The magnetic field also has declination: the angle between magnetic north and true north. Declination varies with location and changes over time. For an animal that wants a stable migratory heading in geographic space, calibration against celestial cues is a sensible strategy. The sky can provide a geographic frame; the magnetic compass can then be aligned to it locally.

Twilight calibration becomes even more important when one remembers that birds often migrate through regions where celestial cues are unreliable. Cloud cover can hide stars for nights at a time. Over water, the horizon is present but can be invisible in darkness and haze. Artificial lighting can wash out the sky. Under those conditions, a magnetic compass is a lifeline. But a lifeline must be attached to something solid. Twilight, because it is a daily event and because it offers multiple structured cues (sun position, polarization patterns, spectral gradients), is one of the most solid anchors the sky provides.

There is a historical irony here. Early naturalists often treated the sun and the stars as the obvious guides for migration. Magnetism was, at best, a later and more exotic suspect. Yet it may be precisely the *combination* of magnetic and celestial cues that gives birds their robustness. A pure star compass would fail on cloudy nights; a pure magnetic compass might drift or be distorted locally. A bird that lets the sky teach its magnetic compass how to behave has something better than either alone: a system that can run on magnetism when it must, but periodically re-ties itself to the broader geometry of the world.

The Emlen funnel, again, plays an unsung role by making such interactions testable. One can expose birds to a particular cue at a particular time and then test the after-effects. That temporal structure is crucial. Calibration is not only about *which* cues are present; it is about *when* they are present. Twilight is a brief window, and experiments that ignore timing can miss the mechanism entirely. The bird might seem indifferent to a cue at midnight, not because the cue is irrelevant, but because the recalibration machinery is not engaged then.

Circadian rhythms enter here as more than a clock-hand. In animals, the circadian system coordinates physiology: hormone secretion, metabolism, attention, sleep pressure. Migration is a state change, not a single behaviour. A nocturnal migrant must be capable of feeding by day and flying by night, a reversal of the ordinary pattern for many small songbirds. The brain regions that process dim visual cues, including those implicated in magnetic compass processing, may therefore be recruited under the circadian umbrella. Cluster N's night-time activation during migration fits naturally into this picture: the brain is not merely awake at night; it is awake in a specialised way.

This raises a delicate point that popular accounts often flatten: calibration is not necessarily conscious. It is not that a bird "decides" to trust the stars today and magnetism tomorrow. The behaviour suggests an underlying set of rules that can be adjusted by experience and context. A bird might be born with a template for how twilight relates to magnetism, then refine that template as it encounters the real sky. Or it might be born with the ability to learn the relationship quickly during a sensitive period. Either way, calibration is a form of learning, even if it is not the kind of learning that looks like training.

One can see why this matters by thinking about the changing world birds now inhabit. The sky itself is being altered by human activity: light pollution changes the visibility of stars and may alter the spectral composition of twilight; atmospheric pollution can change scattering properties; urban environments introduce magnetic noise and distortions. A calibration system that relies on the sky could be strained if the sky's cues are degraded. Conversely, a system that can recalibrate flexibly might cope better with novel conditions, at least up to a point. The scientific details here are still being worked out, but the general principle is sobering: a sensory system evolved under ancient skies is now asked to operate under engineered ones.

A final, small but vivid anecdote returns the calibration story to the scale of an individual animal. In many labs that study orientation, there is a familiar moment after a long night of experiments: the room lights come on, the birds settle, and the paper from the funnel is unrolled. Some mornings the marks are crisp; other mornings they are scattered. When experiments are timed around twilight, the difference can be striking. Birds that seemed uncertain on one schedule may become decisive on another, as if the correct sensory reference has been offered at the correct moment. It is a reminder that navigation is not just about having a compass; it is about keeping it tuned.

A robin leaving at night does not carry a single instrument. It carries a small orchestra of senses that must play in key. The magnetic field provides a steady, global structure. The sky provides a daily calibration signal with a geometry that has guided animals for millions of years. The circadian clock keeps time so that the sun's movement can be interpreted rather than endured. When these systems agree, the bird's chosen direction is not a guess. It is the output of a carefully maintained internal alignment, renewed at dusk and dawn, and strong enough to hold even when the world ahead is dark.

Chapter 6

# Ocean Wanderers and Miniature Migrants

*Navigating the featureless blue void requires a different toolkit than crawling through the clutter of the forest floor. We examine how sea turtles derive global coordinates from invisible field lines to survive ocean crossings, and how the insect world minimizes neural processing to solve complex geometric problems using sensors located in surprising anatomical places.*

## 6.1 Coordinates in the Deep Blue

A hatchling loggerhead sea turtle is small enough to fit in a human palm, yet it enters an ocean that can swallow ships. At the shoreline it hesitates only briefly, then scrambles through the surf and vanishes into water that, to our eyes, contains almost no stable clues. No landmarks. No scent trails that travel for thousands of kilometres. No horizon fixed enough to serve as a reference for more than minutes at a time. Even the sky, if it is visible, changes from hour to hour and can disappear entirely under cloud.

For much of the twentieth century, that early life at sea was simply called the "lost years," because young turtles were rarely seen. The phrase captures a scientific embarrassment as much as a biological mystery: an animal that returns decades later to coastlines and, in many cases, to the neighbourhood of its birth must have solved a navigation problem that defeats most human intuition. The ocean is not a featureless void, but its features are mostly in motion. Currents drift, fronts shift, storms erase patterns. To move with purpose in that world, a turtle needs something both large-scale and reliable.

Earth supplies such a signal. The planet is wrapped in a magnetic field generated deep in the core. It is not a perfectly tidy bar magnet, but it is orderly enough to define, at any point on the surface, a direction and a strength. Two magnetic measurements matter most for navigation.

The first is *intensity*: the total strength of the magnetic field. Intensities vary across the globe, generally increasing toward the poles and decreasing toward the equator, though not in simple concentric circles.

The second is *inclination*: the angle at which magnetic field lines tilt into the Earth. In the Northern Hemisphere, those lines point downward into the ground; in the Southern Hemisphere, upward. At the magnetic equator, inclination is close to zero, with the field more or less horizontal.

A turtle that can sense only a single magnetic feature has, at best, a compass: it can keep a heading. A turtle that can sense *two* features that vary across geography can do something more ambitious. It can estimate where it is, in the same spirit that sailors once estimated latitude from the height of the Sun and longitude from time. The trick is that neither intensity nor inclination gives a unique global address on its own. Along an *isoline* (a contour line of equal value), intensity is the same at many different places. The same is true for inclination. But where an intensity isoline crosses an inclination isoline, their *pair* can become distinctive. Two coordinates, read together, can define a location more precisely than either could alone.

This idea, often called a *bicoordinate magnetic map*, took time to become more than a clever metaphor. For years, the safest claim scientists could make was that sea turtles possessed a magnetic compass. Classic laboratory work in the late twentieth century, associated especially with Kenneth and Catherine Lohmann, showed that hatchling sea turtles respond predictably to changes in the magnetic field, orienting in ways that look purpose-built rather than random. A compass, however, is not a map. A compass tells you which way to go, not where you are. The leap from "turtles sense magnetic north" to "turtles use magnetism as a geographic grid" requires evidence that animals react differently to *fields that correspond to different places*, even when every other cue is stripped away.

That evidence began to accumulate in a particularly elegant form: by bringing the ocean's magnetism into the laboratory.

The key tool is a coil system that can generate controlled magnetic fields. Place a turtle in a water-filled arena inside such coils, and the animal can be exposed, safely and repeatably, to a field that matches what it would experience at some real location. The turtle has never been there; it has no experience to draw on; it cannot smell the water mass or feel the current. Yet it can be asked, in effect: *What does this place mean to you?*

When Nathan Putman and Kenneth Lohmann applied this logic to hatchling loggerheads from Florida, the results implied something far richer than a simple compass. The North Atlantic is organized by a large clockwise circulation, the North Atlantic Gyre, whose moving water can carry small animals far from shore. Loggerhead hatchlings, once in the sea, tend to enter this system: southward and then eastward in warm currents, across to the eastern Atlantic, then back westward and northward, gradually looping toward regions where older juveniles can forage. It is not a perfectly drawn circle; it is a dynamic conveyor whose lanes shift with seasons and storms. But it has a broad structure that can be exploited by an animal that keeps itself within the right band of water.

Now imagine that a young turtle drifts too far north. Cold water threatens, and currents may carry it into ecological trouble. Too far south, and it can be swept into regions that do not connect well to the typical developmental route. In such a life, it is not enough to hold a single heading; the animal benefits from knowing, in a coarse way, whether it has strayed.

In controlled experiments, hatchlings were exposed to magnetic fields corresponding to distinct geographic regions along the gyre. The striking pattern was that turtles did not merely swim in a fixed magnetic direction. They shifted their orientation depending on *which regional field* they were given. A field associated with one part of the Atlantic elicited a response that would, in the ocean, tend to move the turtle toward safer waters; another field elicited a different direction, again consistent with staying within the gyre. The behaviour looked like a set of location-specific "if here, then swim that way" rules, executed by animals at an age when experience is minimal.

Even more provocative was the implication that turtles can extract information analogous to both latitude and longitude. In a human map, latitude is easy: it corresponds to the angle of the Sun or stars, or to the geometry of Earth itself. Longitude historically required clocks; it is the hard coordinate. In magnetism, inclination tends to change strongly with north–south position, making it a plausible magnetic analogue of latitude. Intensity can vary in both north–south and east–west directions, depending on the region. Putman and colleagues showed that hatchling loggerheads could distinguish between magnetic fields from locations far apart east–west but at the same latitude, suggesting that some aspect of the magnetic signature carried usable longitudinal information. For a hatchling with no knowledge of geography in the human sense, the field still seems to contain enough structure to act as a map.

A map, though, need not be perfect. It can be coarse, even blurry, and still be lifesaving. Consider how a person might navigate a dark building by counting doorways and feeling for a staircase: the strategy does not require a precise architectural blueprint; it requires reliable cues that differ between meaningful regions. In the Atlantic, intensity and inclination form broad gradients. Their isolines are not evenly spaced, and they shift slowly over years and decades as Earth's field changes. Yet the intersection of two such gradients can still mark out zones: not points like a GPS coordinate, but areas with recognizably different magnetic "flavours."

The ocean itself helps make such a system workable. Currents create corridors. A turtle does not need to compute an exact position in kilometres if what matters is remaining within, or returning to, a moving stream of warm water that tends to carry it along a productive route. Magnetic information can provide a stabilizing frame, a way to decide when to swim across the flow rather than simply with it.

There is also a human story behind the search for these hidden coordinates. Archie Carr, the influential turtle biologist and writer, helped popularize the idea that sea turtles possess extraordinary navigational abilities, partly by emphasizing their long-distance movements and natal homing. His work inspired generations of researchers to treat turtle migration not as an accidental drifting but as a problem of animal cognition and sensory biology. Yet the details of open-ocean navigation were missing for decades, because the animals themselves were missing. The "lost years" were lost to observers.

Technology began to change that. Satellite tracking and miniature tags, though more useful for larger juveniles and adults than for tiny hatchlings, made it harder to dismiss turtle movements as mere chance. At the same time, laboratory studies became more sophisticated in their manipulation of magnetic fields. These two lines of evidence met in the bicoordinate map hypothesis: turtles do not simply maintain a heading; they appear to *interpret* magnetic signatures as information about where they are within a seascape.

At this point a sceptical question becomes productive: how can a turtle possibly learn a magnetic map if it has never seen a map, never been taught coordinates, and, in the case of hatchlings, never travelled the route before?

One answer is that parts of the map may be inherited, not learned. An animal can be born with a set of responses tuned to particular magnetic combinations if those combinations reliably correspond, in evolutionary time, to regions of danger or opportunity. The hatchling experiments suggest exactly this kind of built-in logic: a magnetic field resembling a risky northern region triggers swimming in a direction that, in the ocean, would tend to bring the animal back toward favourable currents. The turtle need not name the place; it need only react appropriately to its signature.

Inheritance alone cannot solve every navigational task, because Earth's magnetic field is not fixed. It drifts. Over decades, the intensity at a coastline can strengthen or weaken; inclination can shift; isolines can slide relative to land. This secular variation is slow on a human timescale but significant over the lifetime of a sea turtle. A loggerhead that returns to nest twenty or thirty years after hatching is returning to a coast whose magnetic signature has changed since it left.

Rather than being a fatal problem for a magnetic map, this drift becomes a testable prediction. If turtles imprint on, or otherwise memorize, magnetic signatures associated with their natal region, then changes in those signatures over time could nudge where returning adults concentrate their nesting. Putman and colleagues examined long-term nesting records and compared shifts in nesting distributions to changes in coastal magnetic fields. The logic is not that magnetism is the only cue—chemical signals, coastal shapes, wave direction, and perhaps even celestial cues likely contribute—but that magnetism may guide turtles into the right *region* before other senses take over. A coarse magnetic address gets you into the correct neighbourhood; finer cues lead you to the right street.

Such a multi-stage system fits what is known about navigation in many animals. In homing pigeons, for example, different cues dominate at different distances. In sea turtles, the ocean makes this kind of hierarchy almost unavoidable. Chemical cues dissipate; visual cues are absent; currents move the goalposts. A magnetic map, if it exists, is not a luxury. It is the kind of robust, planet-scale scaffold on which other information can be hung.

All of this raises a deeper, and still unsettled, biological question: *what does a magnetic coordinate feel like to a turtle?* Sensing magnetism is not like seeing light or hearing sound. The magnetic field passes through tissue without obvious attenuation; it does not announce itself. Yet animals behave as if they have access to it.

Two broad classes of mechanism dominate discussion. One involves tiny magnetic particles, often magnetite, that could exert mechanical forces on sensory structures as the field changes. Another involves light-dependent chemical reactions in proteins called cryptochromes, where quantum-level effects on electron pairs could make reaction products sensitive to magnetic fields. The second mechanism has been developed most thoroughly for birds, in which magnetic information seems intertwined with vision. In turtles, the sensory and neural hardware remains less clear. The bicoordinate map hypothesis does not require us to know the receptor in order to take the behaviour seriously; sailors used compasses long before they understood electromagnetism. But ignorance of mechanism matters, because a map requires *measurement*: the animal must detect intensity and inclination with enough precision to discriminate regions, and it must do so reliably in the face of noise from waves, temperature, and its own movements.

One way to appreciate the feat is to consider what inclination actually demands. Inclination is an *angle*, a geometrical property of the field vector. Detecting it means comparing vertical and horizontal components of magnetism, not merely sensing "strong" or "weak." Intensity adds another dimension: the *magnitude* of the vector. If a turtle can measure both, it has access to a coordinate pair. If it can also stabilize those measurements while pitching and rolling in the sea, it has solved a sensory engineering problem that would challenge a human instrument maker.

The reward for solving it is a form of navigation that is at once simple and astonishing. The turtle does not need to know the shape of continents. It does not need to calculate currents. It can treat the Atlantic as a set of magnetically labelled zones, each associated with an inherited tendency: swim northeast here, turn south there, hold westward here. This is not a plotted route but a set of rule-based corrections, like a thermostat that keeps a house within a comfort band. The North Atlantic Gyre, with its broad, consistent circulation, makes such corrections meaningful. The turtle's job is not to overpower the ocean; it is to remain coupled to its useful flows and decouple from the dangerous ones.

A human analogy, handled carefully, can help. A smartphone's GPS gives precise coordinates by timing signals from satellites, then computing position with an explicit mathematical model. A turtle is not doing that. Its magnetic map is more like recognizing the *accent* of a place: subtle, distributed, informative in combination. You might not be able to articulate exactly which vowels distinguish Glasgow from Dublin, but you can still respond differently when you hear them. In the same way, a turtle might not "calculate longitude," yet it can behave as if east and west are distinguishable, because the magnetic field carries patterns that correlate with geography.

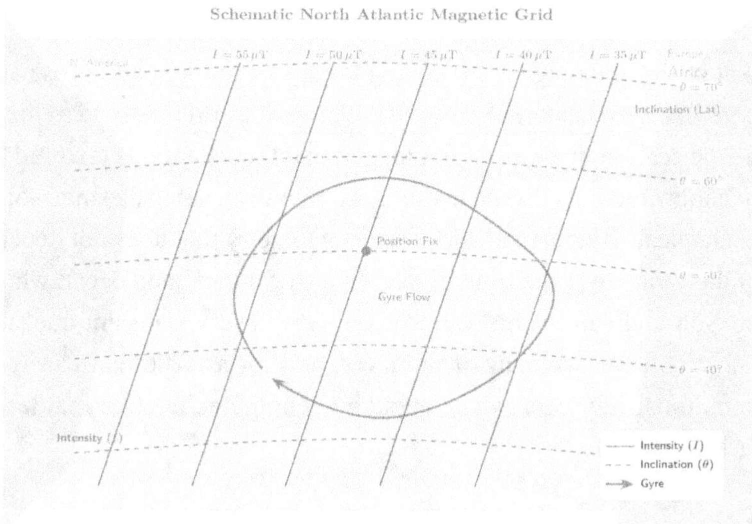


The figure is deliberately schematic, because the point is not cartographic precision but the logic of a grid. Solid lines represent equal intensity; dashed lines represent equal inclination. Neither family of lines, by itself, can tell an animal where it is. A single intensity value occurs along an extended contour; a single inclination value does too. But their crossings produce combinations that are rarer. If a turtle can sense both parameters, those combinations can act like signposts distributed across the ocean. Against a background of moving currents, the signposts are stationary enough to be useful.

The astonishing part is not that Earth has such a grid, but that an animal can exploit it while living in the ocean's restless surface layer. The hatchling's nervous system must connect measurement to action: translate a magnetic signature into a swimming direction that, when combined with currents, keeps the animal within a survivable corridor. This is navigation as feedback control, executed by a creature that begins life with no parental instruction and no second chance. A wrong turn is not a minor error; it can mean cold shock, starvation, or predators.

That makes the magnetic map feel less like a quirky extra sense and more like a foundational piece of the turtle's world. We tend to imagine the sea as empty, but for a loggerhead the Atlantic is textured by invisible gradients. Some textures are chemical, some thermal, some mechanical. Magnetism adds a quiet geometry that does not depend on the weather or the time of day. In a storm-darkened ocean where the Sun and stars vanish and waves erase every transient cue, the planet still offers a pair of numbers, and for a turtle that can read them, those numbers can be enough to choose a direction that leads back toward life.

## 6.2 The Antennae Compass

A monarch butterfly clings to a stem on a cold morning in the American Midwest, wings shut like a book. In a few hours, when the air warms enough for flight, it will lift off and point itself southward. Many will die along the way. The ones that survive reach high volcanic slopes in central Mexico, where oyamel fir forests gather mist and hold a stable, cool microclimate. The butterflies pour into these groves by the millions, turning branches into living drapery.

The unsettling part of the story is not that a delicate insect can fly far, but that it can fly *consistently* far in the correct direction. The autumn migrants are not the same individuals that return north in spring; the full circuit is spread across generations. No monarch is taught the route in the ordinary sense. Yet the fall generation, born in late summer, behaves as if it has inherited an appointment with a particular mountain range it has never seen.

Navigation on this scale forces an animal to face a problem that sailors also face: the world turns. A butterfly that simply keeps the Sun on its left wing will not keep going south all day. As hours pass, the Sun slides across the sky. A fixed angle to the Sun traces a curve, and a curve can mean disaster if the goal is hundreds of kilometres away. The solution is the *time-compensated Sun compass*. In plain terms, the animal uses the Sun as a directional reference but corrects for the time of day. Morning and afternoon require different steering rules because the Sun's position changes predictably.

Time compensation sounds like mathematics, but the butterfly does not compute an equation. It needs a biological clock that keeps time accurately enough to tell the compass how to interpret sunlight. That clock, for a long time, was assumed to live where humans would put it: in the brain.

Monarch research became famous partly because it showed how wrong that assumption can be.

A historical coincidence helped. Monarchs were not always treated as a flagship for insect migration; they became one because their winter colonies were, for decades, a kind of rumour. People in North America watched butterflies stream south each autumn and then vanish. Their overwintering sites were finally located in the mid-1970s, after years of searching, by Fred Urquhart and collaborators, including Mexican naturalists who followed leads on the ground. It is hard to overstate what that discovery did for the science. Once the destination was known, the migration ceased to be a vague marvel and became a testable phenomenon: a route with an endpoint, a schedule, and a set of environmental challenges.

In the following decades, attention turned from *where* monarchs go to *how* they know where to go. The strongest clue, visible even to casual observers, is that butterflies fly on sunny days and often seem confused or less directionally persistent under heavy cloud. That pattern points to a Sun compass. Insects have long been known to use the Sun for orientation; honeybees advertise food directions by encoding the angle of the Sun into their dances. But for a long-distance migrant, the Sun compass must be paired with timing.

The decisive experiments were surprisingly concrete and almost tactile in their simplicity. Christine Merlin and Robert Gegear (2009) manipulated monarch antennae and found that these slender feelers, best known as smell and touch organs, contain circadian clocks essential for proper time-compensated orientation. When antennae were removed, or when their clocks were disrupted, butterflies failed to orient correctly even under a visible Sun. When antennae were intact and functioning, orientation was restored.

The result is counterintuitive in the best way. A butterfly's brain is tiny, but it is still the obvious place to put "computation." Instead, critical timing information is coming from a peripheral structure—a pair of antennae exposed directly to light and the outside world. In an animal whose head is little more than a compact package of sensory and motor circuits, distributing a clock into the antennae may not be a compromise at all. It may be an advantage. Light reaches the antennae without needing to pass through layers of tissue, and each antenna can keep time locally, potentially giving the animal redundancy.

That redundancy became clearer in later work by Patrick Guerra and Merlin (2012). Butterflies can orient correctly with only one antenna providing a properly entrained clock. But if the two antennae are forced into conflict—one exposed to normal light cycles and the other kept in a different light regime—the butterfly becomes disoriented. The most natural interpretation is that the brain receives timing information from both sides and integrates it. When the inputs disagree, the compass loses its reference. A human analogy is a pilot whose two instruments disagree about altitude; it is not the mere presence of instruments that matters, but their agreement.

At this point it is tempting to treat the antennae simply as *clock hands*, supplying time-of-day information for a Sun compass, and to leave it there. But monarchs also sharpen the broader theme of magnetic sensing in a new way: they hint that the antennae are not only timekeepers. They may be part of a magnetic compass system as well.

The magnetic field, by itself, is invisible and silent. For a small flying insect, it could serve as a stable backup when the Sun disappears behind cloud or when a landscape offers few landmarks. A magnetic compass is not expected to replace the Sun compass in monarchs; rather, it could provide a secondary reference, used under conditions where sunlight is unreliable. This cue hierarchy is familiar from many animals. The order matters. A backup cue is easy to miss in experiments if the primary cue remains available. The animal will simply ignore the backup, and the investigator may conclude it does not exist.

Guerra and Gegear (2014) reported evidence that migrating monarchs possess a magnetic compass that depends on light and responds to the *inclination* of the magnetic field lines. Inclination, the angle at which the field points into the ground, provides a north–south sense without requiring the animal to distinguish magnetic north from south by polarity. Many birds show an inclination compass, and it has a signature behavioural property: reverse the inclination and the preferred direction flips, even if the field's polarity remains the same. Monarchs showed responses consistent with such an inclination-based compass.

The dependence on light is more than a technical detail. It points to a family of mechanisms in which magnetism affects chemical reactions triggered by light, often involving cryptochrome proteins. Cryptochromes are found widely in plants and animals and are closely tied to circadian rhythms. Their dual roles—in timing and in light sensing—make them natural candidates for a system in which a magnetic compass is both light-dependent and linked, directly or indirectly, to timekeeping.

A crucial nuance emerged: not just any light will do. The monarch magnetic response appears to require ultraviolet-A and blue wavelengths, roughly in the 380–420 nm range. Without the right spectral band, the magnetic compass can seem to vanish. This helps explain why results across laboratories and years can disagree. An insect tested under lighting that looks bright to human eyes might still be effectively blind to the wavelengths its magnetosensor requires. In the same way, a person can sit in a room flooded with red light and still be unable to distinguish colours properly. The world's "brightness" depends on the biology of the eye—or, in this case, on the biology of the antenna.

Mechanism moved from plausible to persuasive with genetic work. Wan and colleagues (2021) used genome editing to connect monarch magnetic inclination sensing specifically to cryptochrome 1 (CRY1). When CRY1 function was disrupted, butterflies lost the light-dependent magnetic response; CRY2, by contrast, appeared dispensable for that particular behaviour. The details matter because cryptochromes are a family, not a single molecule, and a rigorous claim needs specificity: *which* cryptochrome, in *which* tissue, under *which* light, with *what* behavioural signature.

The antennae return to centre stage here. Both antennae and eyes express CRY1, and both have been implicated as magnetosensory organs. That is not a contradiction so much as a reminder that insects are not built like vertebrates. They do not have a single "retina" that monopolizes light input. Light can be sampled across multiple body parts. An antenna is not merely a smell organ; it is a multifunctional sensory appendage, wired into the central nervous system and exposed directly to the environment. If an organism were to place a light-dependent magnetosensor somewhere, an antenna would be an excellent candidate: thin, translucent, and neurologically connected.

Still, monarchs do not make the broader problem easy, because their migration is guided by many cues. On a clear day, the Sun compass dominates. Under cloud, a magnetic compass might become more relevant. Under scattered cloud, the polarization pattern of the sky can provide directional information even when the Sun itself is obscured. Wind can push a butterfly, and the animal can choose to compensate or drift. Even the landscape can matter: coastlines, mountain ridges, and river valleys can channel flight. The monarch's navigational system is therefore best imagined not as a single "compass needle," but as a decision-making network that weights different signals depending on reliability.

The antennae-based clocks help explain how monarchs maintain the key ingredient for time compensation: a stable internal day. The same biology that makes the Sun compass work also creates a natural interface with light-dependent magnetoreception. A set of molecules that respond to blue light, participate in circadian timing, and can be influenced by magnetic fields sits exactly at the intersection one would predict for an insect that navigates in sunlight and sometimes in its absence.

A concrete example shows why these overlaps matter. Imagine two monarchs leaving the same field on the same October morning. The first has intact antennae. Its internal timing is synchronized to local light cycles, and its Sun compass produces a consistent southwesterly bearing. Clouds roll in by noon. The butterfly loses a clear solar reference, but its antennae still sample ambient light, and its nervous system can fall back on another directional signal that remains stable when the Sun is hidden: magnetic inclination. It may not fly as efficiently, but it avoids turning north or circling aimlessly.

The second butterfly has antennae whose timing information is mismatched, perhaps because one antenna's clock is out of step. Under a bright Sun it may still fly with some directionality, but the steering instructions are subtly wrong because the internal time is wrong. In the afternoon it might drift off course. Under cloud, if the magnetic backup cue also depends on well-defined light conditions and properly functioning cryptochrome pathways, the butterfly is doubly compromised: it has lost both its primary reference and the reliability of its backup.

That scenario is not just a thought experiment. It captures a real biological point: navigation systems are often *fragile in specific ways*. A small change in spectral composition of light can erase a magnetic response. A conflict between two peripheral clocks can scramble a Sun compass. The fragility is a clue to the underlying mechanism. It is also a reminder that migration is a chain of dependencies; break one link and the whole behaviour can degrade.

A broader caution is needed, because it is easy to overgeneralize from monarchs to all insects. Not all magnetic behaviours in insects appear to fit the same light-dependent, inclination-based template. Work on desert ants of the genus *Cataglyphis* reported polarity-sensitive magnetic orientation during early learning behaviour, implying a mechanism that does not sit comfortably within a pure cryptochrome radical-pair inclination compass. A polarity-sensitive compass distinguishes magnetic north from south by sign, not just by tilt, and that behavioural difference matters. It suggests that insects may use multiple magnetic sensors or multiple strategies, even within the same broad group of animals.

This diversity should not dilute the monarch story. It makes it more interesting. The monarch is not an all-purpose model of magnetoreception; it is a clear example of a particular solution: a light-dependent magnetic cue embedded in a migratory toolkit that is already built around light and time. The antennae are the perfect anatomical symbol of that solution. They are both exposed to the sky and intimately wired into the brain's orientation circuitry. They host clocks. They host light-sensitive proteins. They are paired, allowing bilateral comparison and redundancy. They are, in a literal sense, the butterfly's forward-facing feelers for an invisible geometry.

There is a final human resonance that tends to be overlooked when the focus stays purely on mechanism. Monarch migration is increasingly precarious. The oyamel fir forests that shelter overwintering colonies face logging pressures and climate-related stresses. Breeding habitats in North America have changed under agriculture and urban development, and the availability of milkweed—the caterpillar's host plant—has been altered in many regions. A navigational system can be exquisitely tuned and still fail if the places it expects no longer exist in the same form.

Against that background, the antennae compass feels less like a laboratory curiosity and more like a fragile inheritance, an evolved relationship between an insect, the sky, and a planet-sized field. A monarch on the wing carries a pair of antennae that are at once clocks, light sensors, and, at least under the right wavelengths, a gateway to magnetic direction. When the Sun is sharp, the insect can steer by daylight and internal time. When daylight becomes unreliable, the same light-sensitive machinery can keep a different kind of bearing, drawn not on a map but in the tilt of Earth's field lines beneath its wings.

## 6.3 Rolling Straight and Building True

A dung beetle at a fresh dung pile is surrounded by chaos. Flies arrive first, then beetles in numbers that can make the ground seem to move. Competition is immediate and physical: a good scrap of dung is both food and a future nursery, and the quickest animal wins. Many ball-rolling species solve the problem with a blunt strategy. They carve out a piece, shape it into a ball, climb on top, and roll it away as fast as possible—often backwards, pushing with their hind legs while steering with the front.

The urgency makes the navigational demand unusually clear. The beetle does not need to find a hidden overwintering forest or cross an ocean. It needs a *straight line* for tens of metres. Straightness is not aesthetic; it is survival. A meandering path keeps the beetle within sight and scent range of rivals who can steal the ball. A direct path quickly moves the prize into quieter ground where burial becomes possible.

This is a different scale from the open ocean or a continent-spanning migration, but it tests the same underlying idea: orientation is not a single sense so much as an ability to lock behaviour to a stable reference. For the beetle, the reference is usually celestial. Even when the animal rolls backwards, it can keep a constant bearing by fixing its internal compass to a cue in the sky. The technical term often used for this is *menotaxis*: maintaining a constant angle relative to a distant cue, rather than aiming directly at a goal.

What makes dung beetles compelling is that their "distant cue" can be astonishingly flexible. On a bright day it can be the Sun. At dusk it can be the pattern of polarized light in the sky, an invisible geometry created when sunlight scatters through the atmosphere. On moonlit nights the Moon itself can serve. Some species can even use the diffuse glow of the Milky Way as an orientation reference. The practical lesson is that an internal compass does not need a single needle; it needs a way to represent direction and update it reliably as the animal moves.

Behavioural biologists have made this visible in a way that is almost theatrical. A beetle rolls a ball away from a dung pile in a clean line. The experimenter quietly rotates the sky cue—by shifting a light source, or by placing the animal under conditions where the polarization pattern is altered—and the beetle's path turns accordingly. The beetle behaves as if its world contains a coordinate frame painted on the sky, and it is keeping its motion locked to that frame.

There is a revealing ritual that many ball-rollers perform before setting off. They climb on top of the ball and execute a brief "dance": rotating, pausing, rotating again. It looks like indecision until one realizes what it accomplishes. While turning on the spot, the beetle can sample the sky from multiple orientations and calibrate a heading. Once rolling begins, the beetle often stops less frequently, as if it has set a reference and is now protecting it from drift. Even without any mathematics in the human sense, the behaviour embodies a principle familiar to engineers: measure carefully at the start, then maintain the set point.

This story might sound purely celestial, but magnetism is never far away, even in a creature that seems to take its guidance from the heavens. Magnetic cues can contribute in at least two ways. First, Earth's field is available when the sky is compromised—by cloud, vegetation, or the animal's own posture. A beetle pushing a ball backwards spends much of its time with its eyes oriented toward the ground and the horizon broken by grasses or uneven terrain. Second, magnetic information can provide a stable baseline for calibration. If the sky cue changes abruptly, a magnetic reference can prevent a catastrophic reorientation.

Dung beetles are therefore a useful reminder that magnetic sensing does not always appear as an animal "following magnetism." Often it appears as *robustness*. A system that uses multiple cues can keep operating when one cue becomes noisy. That is hard to spot unless experiments deliberately remove the dominant cue.

In insect navigation, a small brain does not imply a simple solution. A major reason is that insect brains contain specialized circuitry for direction, often centred on a region called the *central complex*. This is not a metaphorical "centre," but a real set of interconnected structures that functions as an internal compass. Neurons in this circuit can represent heading as a kind of activity bump that rotates as the animal turns, allowing it to keep track of direction over time. This is the same neural logic that helps a beetle roll straight, a bee fly a stable course, or an ant return to its nest.

Once such a circuit exists, the sensory problem becomes one of feeding it the right reference signals. A polarization pattern in the sky, the position of the Sun, the gradient of light intensity, and potentially magnetic cues can all be translated into inputs that stabilize the internal compass. Different cues can be weighted depending on conditions. Work on dung beetles has shown that cue reliability changes with solar elevation: at certain times of day the polarization signal is strong and precise; at others, the Sun's position may be a better guide. The beetle's behaviour suggests it can switch emphasis without needing to "know" the physics—it simply relies more heavily on what is most consistent.

It is easy to admire the elegance and forget the messy context in which this evolved. Rolling a ball backward over uneven ground is a sensorimotor nightmare. The beetle must steer while its body orientation is partly decoupled from its travel direction. Any internal compass must be updated not only by visual rotation but by the animal's own movements, slips, and bumps. That this works at all hints that the compass circuit is not a delicate instrument. It is built to tolerate noise, and it likely draws on whatever stable reference is available. Magnetism, even if not always the primary cue, offers exactly the kind of background stability that helps a noisy system behave cleanly.

A second, very different, case brings magnetism into the foreground not as a compass for motion but as an alignment guide for construction. Social insects build. They build in darkness, underground, inside trees, in cavities of stone. Their structures have geometry: combs in bees, galleries in ants and termites, nest entrances aligned in particular ways. Geometry implies a reference frame. Humans reach for a level, a compass, a string line. An insect has none of these, yet produces striking regularity.

Honeybees, for example, build comb with a consistent orientation of cells, and they can do this in conditions where visual cues are minimal. Experiments have suggested that, in total darkness, bees can use Earth's magnetic field as a directional reference. The logic is straightforward. If an insect can sense magnetic direction, it can align construction relative to that direction even when sunlight and landmarks are absent. A magnetic cue is especially plausible inside a cavity because it penetrates wood and wax. It does not need a line of sight.

The bee's comb is a reminder that navigation is not always about going from A to B. It can be about maintaining a shared coordinate system among many individuals. A colony is a distributed builder. No single bee holds the blueprint. Consistent orientation across the work of thousands of workers requires a common reference—a way for each bee's local decisions to add up to a coherent architecture. Whether magnetism is the dominant reference in real nests is still debated, because bees also have access to gravity, airflow, and subtle temperature gradients. But the very plausibility of magnetic alignment in darkness underlines why magnetoreception is useful even at small spatial scales: it provides an omnipresent axis.

Termites push this idea to a grand, and controversial, extreme. Some termite species build tall mounds with a pronounced orientation: flattened like a blade, with the long axis aligned roughly north—south. Early observers found this alignment so consistent in some regions that the mounds earned the nickname “magnetic termite mounds.” The phrase is irresistible: it suggests that termites are literally building with a magnetic compass in hand, aligning their cities to the meridian.

The scientific story is more interesting than the nickname. North—south alignment can be explained without invoking magnetism at all. A mound is not simply a pile of soil; it is a climate-control device. By presenting a narrow face to the harsh midday Sun and broader faces to morning and afternoon sunlight, a north—south oriented “blade” can moderate internal temperature swings. Wind direction, shade, and the daily rhythm of heating can all favour certain orientations. Later analyses argued that variation in mound orientation across landscapes correlates better with thermoregulatory optimization shaped by local winds and solar exposure than with a simple magnetic template. In other words, the mounds may look “magnetic” because the best engineering solution often aligns with geographic north—south for reasons of heat and airflow.

This is not a defeat for magnetoreception; it is a cautionary tale about inference. When an animal structure aligns with the cardinal directions, magnetism is an obvious suspect, but it is not the only one. The cardinal directions also structure the Sun’s path, prevailing winds, and temperature gradients. A pattern that appears to “match the magnetic field” can arise because it matches the thermal environment. To claim magnetism, one needs experiments that separate these cues—for instance, by altering the magnetic field while leaving sunlight and wind unchanged, or by showing that animals maintain alignment in darkness where solar cues are absent.

At the same time, the termite case should not be dismissed as mere misnaming. Even if thermoregulation explains the mound's final orientation, termites still face a question that parallels the turtle's ocean and the monarch's sky: how do you keep a stable reference amid noise? A builder operating over weeks in variable weather could benefit from anchoring construction to a cue that does not change with clouds or seasons. Magnetism offers that anchor. It is entirely possible that termites, bees, and ants use magnetic information at some stages—perhaps to keep galleries straight underground, or to stabilize orientation during brief exposures to the surface—even if the mature mound's shape is ultimately tuned by temperature and wind.

Ants make this interplay between cues vivid because they switch between them so fluidly. Desert ants, in particular, are famous for *path integration*: keeping track of distance and direction travelled so they can return to a nest entrance that is nearly invisible. They do this using the Sun and polarized skylight, and they also use landmarks when available. Recent work has added a twist by suggesting that some ants can use magnetic information in a way that is sensitive to polarity. Polarity sensitivity means the animal distinguishes magnetic north from south as *opposites*, rather than merely using the inclination angle. That behavioural signature matters because many light-dependent magnetoreception mechanisms are expected to produce an inclination compass, not a polarity compass. A polarity-sensitive response points either to a different mechanism or to a more complex system than a single sensor.

The practical meaning is simple: insects are not committed to one magnetic solution. Just as birds, turtles, and fish may employ different magnetosensors or use magnetic information in different ways, insects can too. A monarch's light-dependent inclination compass, tied to cryptochrome and expressed under specific wavelengths, may coexist in the animal kingdom with an ant's polarity-sensitive orientation that hints at other receptors. Diversity here is not an annoyance. It is a clue that evolution repeatedly discovered that magnetism is too useful to ignore.

One can see a common thread across these small-scale stories without forcing them into the same mould. A turtle in mid-ocean uses magnetic parameters to interpret its position in a shifting current system. A monarch uses light and internal time, with magnetism as a stabilizing backup when the sky becomes unreliable. A dung beetle uses the sky to keep a straight line away from intense competition, but it benefits from a navigation circuit that can integrate multiple references and maintain a heading under noisy movement. A bee or termite colony constructs with consistent geometry, implying a shared reference frame that could, in certain contexts, be magnetic.

The shared thread is not distance; it is *geometry*. Magnetism is one of the few environmental signals that provides geometry everywhere on Earth. It is global, penetrative, and largely indifferent to weather and time of day. Once an animal has any capacity to sense it, the signal can be used in many ways: as a heading reference, as a positional label, as a calibration tool, as a scaffold for collective building.

A last image brings the argument back to ground level. In the late afternoon, the dung pile has cooled and the frenzy has eased. A beetle that escaped early has buried its ball beneath the soil. The surface looks undisturbed, as if nothing happened. Yet under the ground a larva will develop inside that buried sphere of dung, protected from drying air and from thieves. A straight path was not merely a line on sand; it was the difference between losing everything and building a future. In the life of a small insect, the abstract idea of a compass becomes concrete: a neural commitment to a direction, sustained long enough for an animal to turn a moment of chaos into a hidden, carefully aligned refuge.

## Chapter 7

# Setting the Dial: Development and Experience

*A biological compass is rarely ready-made at birth; it requires precise calibration to be useful. This chapter uncovers how hatchlings and fledglings transform raw magnetic data into a functional map, using the rotation of the stars to define 'north' and memorizing the magnetic coordinates of their birthplace to ensure a return years later.*

## 7.1 The Magnetic Signature of Home

A green sea turtle the size of a dinner plate slips into the surf and vanishes, as if it had never existed. For a few frantic minutes it is all flippers and foam, dodging crabs, gulls, and the pull of the backwash. Then it becomes something more like a planktonic leaf, drifting out beyond the breakers into a blue expanse where landmarks dissolve. If it survives the first hours, it will spend years in a world that is, to human senses, nearly featureless. And yet the adult that returns decades later will often do something astonishingly specific: she will climb ashore not merely on the right coastline, but on a narrow ribbon of sand that is, for her species, *home*.

Salmon perform a parallel feat in a different medium. A juvenile salmon leaves its natal river, is swept into an ocean that is vast even by animal standards, and later reappears as an adult with gonads ripened for a single spawning attempt. It finds not just the correct continent, or the correct sea, but the correct river system, and then pushes upriver toward a particular tributary. Smell plays a celebrated role in the last stages of this journey, when river water carries an intricate chemical signature. But chemical cues cannot reach across ocean basins. Something else must get the animal close enough for smell to take over.

The guiding idea here is deceptively simple: young animals may *imprint* on the Earth's magnetic field at birth (or soon after), storing a "magnetic signature" of their natal area. Much later, from far away, they can use magnetic information again—not just as a compass direction, but as a kind of address—to relocate the region where they began.

Imprinting is a familiar concept in animal behavior. Konrad Lorenz's famous work in the 1930s made "imprinting" almost a household word: goslings that saw Lorenz shortly after hatching treated him as a parent and followed him faithfully. In many species, early experience creates a lasting preference or memory, locked in during a sensitive period. Natal homing can be thought of as imprinting stretched across geography: a young animal forms a memory of *where* it is, then later seeks that place out.

But "where" is a slippery notion in the open ocean. If the Earth were a featureless sphere, and if animals could not carry maps, then "where" could only mean "in some direction." The magnetic imprinting hypothesis makes "where" more concrete by borrowing a cartographic trick. A point on a map can be specified by two coordinates: latitude and longitude. The magnetic field also varies across the planet in ways that can, in principle, supply two coordinates. You do not need to know latitude and longitude as numbers; you only need to recognize a particular combination of cues as matching what you once experienced.

Two magnetic features are especially useful in this role. The first is *inclination angle*: the tilt of the field lines relative to the Earth's surface. Near the magnetic equator the field is roughly horizontal; nearer the magnetic poles it plunges more steeply into the ground. The second is *intensity*, the field's strength, which also changes geographically. Each of these cues forms a set of contour lines on the globe: isolines of equal inclination, and isolines of equal intensity. Where one inclination isoline crosses one intensity isoline, the intersection marks a particular region. If an animal can detect both cues, then "home" can be encoded as a distinctive crossing of two magnetic contours: a bi-coordinate address.

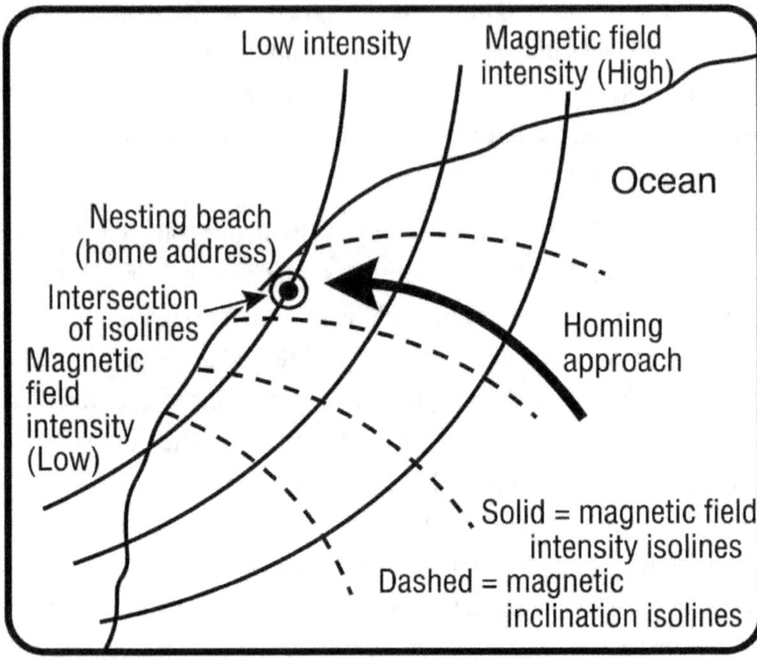


Conceptual map: a unique intersection can mark a region even without landmarks.

The diagram is intentionally idealized. Real magnetic isolines do not form tidy, evenly spaced curves, and coastlines are not obliging. But the logic matters more than the aesthetics. A single magnetic cue is ambiguous. If you only knew the inclination angle you were seeking, you could be anywhere along a broad band that wraps around the planet. Add intensity, and the uncertainty shrinks to the neighborhood where the two bands cross. For a homing animal, "neighborhood" can be good enough. Once the traveler reaches familiar regional waters, other senses and learned geography—smell, currents, wave direction, coastal shape, even the soundscape of surf—can refine the approach.

This is also a good place to be honest about scale. Magnetic imprinting is not a claim that a turtle returns to the exact square meter of sand by magnetism alone, like a guided missile. It is a claim about *getting into the right postal district.* Many nesting beaches are long and continuous; many rivers have branching tributaries. Precision can be layered: a magnetic map to reach the right stretch of coastline, then local cues to choose the right beach access, and finally tactile and visual cues to dig a nest above the tide line.

The question is whether the field provides usable information and whether animals actually use it. Experiments and observations over the last few decades have accumulated into a picture that is persuasive without being simplistic.

A key distinction in animal navigation is between a *compass* and a *map.* A compass tells you which way to go; a map tells you where you are. Magnetic compass sense—the ability to maintain a heading relative to the field—is now well supported in many animals. A magnetic map is a stronger claim: it requires animals to detect at least one feature of the field that changes with location, and to interpret that feature as positional information. The imprinting hypothesis takes the further step of saying that part of that map is learned early: the animal stores a reference pair of magnetic values associated with its natal area.

One of the most compelling forms of evidence comes not from single animals in a laboratory tank, but from populations across decades. The Earth's magnetic field is not fixed. It drifts and warps slowly as molten iron moves in the outer core. That slow reshaping is called *secular variation*, and it creates an opportunity for a kind of natural experiment. If animals truly use magnetic signatures as targets, then changes in magnetic contours over time should leave fingerprints on where animals choose to go.

In 2015, J. R. Brothers and Kenneth Lohmann analyzed long-term nesting records for loggerhead sea turtles along the Florida coastline and compared them with how the local geomagnetic field had changed. The geomagnetic imprinting model makes a clear prediction: if two nearby sites become magnetically more similar over time (their signatures "converge"), turtles should have more trouble distinguishing them and nesting should spread or shift accordingly; if signatures become more distinct ("diverge"), nesting should become more concentrated in the magnetically distinctive areas. The remarkable point is not that every nest location can be predicted—nature is too messy for that—but that the broad patterns of change aligned with the magnetic changes in ways the hypothesis anticipates. The field's slow drift, usually an inconvenience, becomes a clock hand that lets us test a behavioral mechanism.

Salmon offer a different lens on the same idea because their migration is not merely a return to a coastline but a directed movement through oceanic regions. Nathan Putman and Kenneth Lohmann have argued that geomagnetic imprinting can guide salmon to the general area of their natal river before smell takes over in the final approach. A neat feature of this proposal is that it helps reconcile two observations that otherwise sit awkwardly together: salmon are exquisitely sensitive to chemical cues in river water, but they also show large-scale consistency in marine routes that are hard to explain with olfaction alone. If a juvenile imprints on a magnetic signature near its river mouth or on the continental shelf, then an adult could seek that same magnetic combination years later, reaching the correct coastal region even when chemical signals are unavailable.

A historical anecdote can sharpen the intuition here. Long before anyone spoke of "magnetic maps," salmon were already teaching humans about memory and return. In the late nineteenth and early twentieth centuries, fisheries scientists began to mark fish and study their movements, hoping to manage stocks that were economically and culturally vital. Tagging programs revealed that salmon often returned with striking fidelity to their natal waters. That fidelity was a biological marvel and a management headache: it meant that "salmon" was not one uniform resource but a mosaic of local populations, each with its own destiny. The imprinting idea has always hovered in the background of such work. What has changed is that magnetism offers a physical, measurable candidate for the long-range part of the puzzle, one that can be mapped and that changes in predictable ways over time.

The idea of a "magnetic signature" also meshes with a second kind of evidence: population structure. If animals use magnetic cues to home, then the magnetic landscape should shape which groups interbreed and which remain distinct. In 2018, a study testing magnetic navigation and geomagnetic imprinting in sea turtles linked spatial genetic patterns to features of the geomagnetic field, suggesting that where the magnetic field provides clearer, more distinguishable signatures, populations may be more strongly separated. Genetics cannot tell you what an individual turtle sensed on a given night, but it can reveal the long-term consequences of navigation strategies. A homing mechanism that is even modestly reliable will, over generations, carve boundaries into the gene pool.

At this point a skeptic might object that magnetic values are not unique. Many places on Earth share similar intensities or inclinations, and isolines can run roughly parallel along long coastlines. Would that not make magnetism a poor address system? The answer is: it depends where you are and what scale you need. Along some coasts, intensity and inclination isolines meet the shoreline at angles that create distinctive crossings for different nesting regions. Along other coasts, the magnetic "grid" can be stretched, making nearby locations magnetically similar. The imprinting hypothesis therefore carries a built-in prediction: homing accuracy should vary geographically with the geometry of the magnetic field. Where the field creates sharper contrasts over short distances, magnetic homing should be easier; where contrasts are shallow, it should be harder and animals should rely more heavily on other cues.

There is also the question of timing. When, exactly, would a turtle or salmon "take the magnetic snapshot" of home? Imprinting implies a sensitive period, a window during which the nervous system is primed to store a durable association. For salmon, a natural candidate is the smolt stage, when juveniles undergo physiological changes that prepare them for salt water and begin their seaward migration. For turtles, candidates include the time on the natal beach, the first hours in nearshore waters, or early life in coastal nursery regions. The details are likely to vary by species and may not be a single moment so much as an interval during which the magnetic environment is sampled and stabilized into memory.

One reason this is hard to pin down is practical: it is easier to bring a young bird into a planetarium than to follow a baby turtle into the Gulf Stream. Yet the imprinting concept is not mystical; it makes tractable demands on biology. The animal must be able to detect magnetic inclination and intensity with sufficient resolution, and it must have neural machinery that can store and later recognize a match. Both are plausible. Magnetic detection mechanisms discussed elsewhere—magnetite-based receptors that respond to field strength, and light-dependent radical-pair chemistry in cryptochromes that can encode directional information—give animals multiple "handles" on the field. The open question is not whether magnetism can be sensed, but how sensory signals are combined, remembered, and used as part of decision-making.

A useful way to think about the memory problem is to compare it with how humans remember places without coordinates. A person who grew up in a neighborhood can return years later by recalling a *constellation of cues*: the smell near a bakery, the sound of a particular intersection, the slope of a street, the way light falls in late afternoon. None of these cues is unique in isolation. Together they form a recognizably "right" mixture. A magnetic signature could function similarly: not a single number, but a learned combination of magnetic conditions that, when encountered again, triggers the "this is it" recognition that biases movement and searching.

The imprinting hypothesis also accommodates an important psychological fact about navigation: it is often hierarchical. When people look for a friend's apartment, they first get to the right city, then the right neighborhood, then the right street, and only then do they scan door numbers. Animals can do the same without words. A turtle returning to a coast might first aim for a broad latitudinal band (inclination), then refine position by intensity, then follow coastline shape, currents, and finally local beach cues. A salmon might first reach the correct continental margin, then the correct region of the shelf, then switch to odor cues that guide it into its natal river system.

The power of this framework is that it makes "home" a physical target that moves slowly over time. That movement is not merely a complication; it is a test. If the field's contours drift, then an animal returning decades later is not trying to match the field as it was on its birth night exactly. It is trying to match a memory that may now correspond to a slightly displaced region. The animal's solution could take several forms: tolerate some mismatch and rely on other cues; update the target during early life or during intermediate returns; or build in a rule that treats magnetic signatures as approximate, not absolute. In practice, a small mismatch might simply widen the search pattern along a coast, leading to subtle shifts in nesting density from year to year. When those subtle shifts line up with the direction and magnitude of geomagnetic change, as in the loggerhead analyses, the case for magnetic imprinting becomes much more than a clever story.

A final, sobering twist is that the same mechanism that helps animals find home can also, in a changing world, become a trap. Coastlines erode, rivers are dammed, artificial lights lure hatchlings inland, and ocean temperatures reshape feeding grounds. If a turtle's internal "address" brings her faithfully back to a beach that is now developed or flood-prone, fidelity becomes vulnerability. Nature built a navigation system for a planet whose magnetic field and coastlines change gradually. Human alterations can be abrupt. Understanding magnetic imprinting is therefore not only about satisfying curiosity. It changes how conservationists interpret where animals *should* go, where they *try* to go, and why protecting a species may require protecting not just feeding grounds and migration corridors, but the particular natal sites that anchor an animal's lifelong sense of place.

To stand on a nesting beach at night is to feel how narrow that anchoring can be. The ocean is immense; the target is a strip of sand. A female turtle, heavy with eggs, drags herself above the tideline and begins to dig with slow, deliberate strokes. Beneath her, the Earth's field passes silently through water, shell, and bone, offering no scent, no shimmer, no sound—just a structure. If she carries a magnetic memory from her first hours of life, then this beach is not simply where she happens to arrive. It is where an old internal match finally clicks into place, and the abstraction of a planet-wide field becomes, for one animal on one night, the unmistakable feeling of being home.

## 7.2 Starlight and the Reference System

On a clear autumn evening a small songbird can make a decision that looks, from the ground, like a flick of fate. It feeds in a hedgerow, grows restless as the light thins, then launches into darkness and aims into a sky that seems empty. Within minutes it is beyond the range of any landmark a bird could have learned in daylight. Hours later it may be crossing country, then sea. The feat is not simply endurance. It is direction held steady when the world offers almost nothing solid to hold onto.

A magnetic compass can supply a heading, but a heading is only as good as the reference frame that defines it. Human compasses are calibrated to a convention: north is marked on a map, and we learn early that "north" is not "up" in the sky or "toward a mountain," but a global direction. A bird does not inherit our conventions. It can inherit a magnetosensory system, and perhaps an instinct to depart in a seasonally appropriate direction, yet still need to answer a quieter question: *relative to what*? When the magnetic field says "this way," how does the animal know what that "this way" corresponds to in the larger world?

The Earth provides more than one stable scaffold. There is the Sun's arc, which offers a time-compensated compass if the animal knows the hour. There is the night sky's pattern of stars, which seems fixed from one night to the next. And there is something even more reliable than any single star: the slow turning of the entire canopy around a point. The stars wheel as the Earth rotates, tracing circles centered on the axis of rotation. To a human observer in the Northern Hemisphere that center lies close to Polaris, the North Star. To a young bird, the crucial information is not a named star, but a geometrical fact: the sky rotates around a stable pivot.

A deep insight of twentieth-century behavioral biology is that a bird's directional systems are not isolated gadgets—one for magnetism, one for stars, one for the Sun—but a coordinated set that needs calibration. Calibration is the act of aligning instruments so that they agree on what "north" and "south" mean, or more precisely, so that a preferred migratory direction is anchored to the external world. Without calibration, different compasses can contradict one another, and the animal can do everything "right" while still departing in the wrong direction.

The most famous demonstrations of celestial calibration came from a deceptively simple device: a funnel. Stephen Emlen, working in the late 1960s and publishing a landmark study in 1970, built what became known as the Emlen funnel to measure migratory restlessness and orientation in captive birds. The set-up was humane and ingenious. A bird placed in a slanted-walled funnel, during the season when it would normally migrate, becomes active and attempts to take off. Its hops and scrabbling leave marks on the paper lining the walls. Those marks, when counted, reveal a preferred direction even when the bird cannot actually fly away. The funnel transformed an invisible intention into data.

Emlen's key move was to place birds under a planetarium sky—a dome where the star pattern could be controlled like a laboratory variable. Under a natural sky, a migratory bird can see the whole rotating star field. Under a planetarium, the researcher can speed up, slow down, rotate, or even invent a sky that could not exist outdoors. This is where the story stops being romantic and becomes almost philosophical. If you can teach a bird a fictitious night sky, you can discover what, exactly, it is learning.

The results were startling. Young birds that had not previously seen the night sky did not show normal orientation. But when they were allowed to watch a rotating star pattern, they began to orient in seasonally appropriate directions. More tellingly, if the apparent axis of rotation was shifted—as if the world were turning around a different pole—the birds oriented according to that learned axis. They were not born with a fixed "star map" that matched particular constellations to particular directions. They were learning a reference system from rotation itself. The sky, to them, was not a picture but a moving geometry.

It is easy to underestimate what a radical conclusion this was. Many instincts are remarkably specific. Young sea turtles, for example, can show appropriate offshore headings with no training; many birds inherit migratory direction preferences. Yet here was an element that required experience: an anchoring of direction to the true rotational symmetry of the world. Emlen's work did not claim that birds learn everything. It showed that a crucial piece of the navigational apparatus—the *coordinate frame*—could be acquired in a developmental window.

How does this connect to magnetism? The magnetic compass, unlike a human compass needle, does not necessarily provide a labeled "north" in all species. Many birds have an inclination compass: it distinguishes "poleward" from "equatorward" by reading whether field lines tilt down or up, rather than by detecting polarity. In an inclination compass, a field with lines slanting down in front might be treated as the appropriate "northward" direction, but the system must still be assigned a meaning. If a bird's magnetic compass can, under some conditions, confuse north and south, then celestial information can provide a fix: it can tell the animal which end of the axis corresponds to geographic north.

Even for a polarity-sensitive compass, there is another complication. Magnetic north is not the same as geographic north across much of the globe. The angular difference between them is called *declination*. Declination varies by location and changes over time. A navigator that uses magnetic directions but wants to maintain a consistent relationship to geography must either tolerate this mismatch or correct for it. Humans solve this with maps marked with declination, or by converting between "magnetic" and "true" bearings. Birds cannot consult a chart, but they can perform a biological equivalent: repeatedly calibrate the magnetic compass using celestial references that track true geographic directions.

Evidence that birds perform such calibration came from experiments beyond the planetarium. In 1990, Kenneth and Margaret Able provided results suggesting that celestial rotation could calibrate the magnetic compass of migratory birds during development. The essence of these experiments is a cue-conflict design: present magnetic information and celestial information in a way that can be experimentally shifted, then observe which cue controls the bird's subsequent orientation. When the celestial rotation reference is altered, the bird's magnetic orientation can shift accordingly, as if the bird has used the sky to set the zero point of its magnetic dial.

A calibration story, however, needs a mechanism that operates in the real world, not just in carefully staged night-sky lessons. Stars are not always visible. Clouds come and go; the horizon can be blocked by vegetation; many migrants travel on nights with partial cloud cover. If calibration required pristine star visibility every evening, it would be fragile.

A more robust cue exists at the edges of day. At sunrise and sunset, the sky contains a pattern that most people never notice: polarized light. Light waves can vibrate in many planes; polarization describes a bias toward one plane. Scattering in the atmosphere produces a predictable pattern of polarization across the sky, especially near the horizon during twilight. To many animals, including birds, this pattern is not subtle. Their eyes can detect polarization, and the pattern provides compass information even when the Sun itself is not visible.

In 2006, Roswitha Muheim and John Phillips reported in *Science* that polarized light cues at twilight can underlie magnetic compass calibration in migratory songbirds. The implication is powerful. Twilight occurs every day. Even under clouds, the polarization pattern can sometimes be detectable in patches of sky. And twilight is exactly the time when many nocturnal migrants are preparing to depart. A bird can check the sky's polarization pattern as a daily reference, then set its magnetic compass accordingly for the night's flight.

The idea of daily recalibration is supported by field telemetry work, including studies associated with Wolfgang Wiltschko, Henrik Mouritsen, and colleagues, and by experiments where birds are exposed to altered magnetic fields during the twilight period. The pattern that emerges is subtle but consistent: manipulations at twilight can shift subsequent departure directions, while the effect may be temporary, resetting after the birds experience normal twilight again. That is the signature of calibration rather than permanent learning: a compass being adjusted, not rewritten.

Calibration also explains why magnetoreception can look capricious in experiments. Sometimes a study finds strong magnetic orientation; another finds none. Sometimes birds follow magnetic cues; sometimes they seem to ignore them. Cue access matters. A bird indoors with no view of the horizon may lose polarization information and thus lose the ability to calibrate that day. A bird under an artificial sky may see stars but not the proper rotation pattern. A bird tested at midday may rely on a Sun compass instead. Species differ as well: some are more reliant on magnetic cues, others more on celestial cues, and many flexibly switch. Navigation is not a single algorithm; it is a negotiation among imperfect instruments.

A concrete image helps here. Imagine a sailor with multiple tools: a compass, a chronometer, and a view of the stars. On a clear night the sailor checks Polaris and adjusts the compass's deviation. On a cloudy night the sailor trusts the compass more, but with the memory of the last clear calibration. The point is not that any one tool is always in charge, but that agreement among tools reduces accumulated error. A migratory bird, operating with a brain the size of a thumbnail, appears to do something functionally similar.

The most counterintuitive part of the story is that the calibration cue is not necessarily a "north arrow" in the sky. Emlen's planetarium birds were not memorizing "the Big Dipper points to north." They were learning the axis around which the star field rotates. That axis is defined by physics, not by culture. In the Northern Hemisphere it sits near Polaris; in the Southern Hemisphere, where there is no bright pole star, the axis is still there, marked only by the rotation of constellations. A reference based on rotation works everywhere.

The emphasis on rotation also shows why calibration must have a developmental component. A naive juvenile needs to learn what the world's rotation looks like from its latitude and its view of the sky. The same star can rise and set differently depending on where you are. Trees and hills can block parts of the sky. A bird does not need to learn every star, but it does need to learn which nightly motion counts as "the sky turning" rather than "a moving cloud pattern." Once learned, that rotation axis becomes an abstract anchor that can be used again and again.

One might wonder whether the magnetic compass itself requires learning. Many experiments suggest that the basic ability to orient magnetically is present early, but the alignment and integration with other cues is what experience tunes. "Tuning" is not a metaphor here; it is an engineering description. A sensor can be functional yet misaligned. A factory-fresh compass can point, but if the housing is mounted a few degrees off, every bearing will be biased. For a migrant crossing a continent, a bias of even a few degrees can matter.

It is tempting to treat "innate" and "learned" as opposites, but migratory navigation dissolves that dichotomy. The bird inherits sensory hardware and motivational programs, and it inherits some directional preferences. Then the world supplies a calibration signal that makes those programs reliable. The most elegant feature of the calibration approach is that it uses global regularities—celestial rotation and twilight polarization—to tune another global cue, the magnetic field.

Those regularities were stable long before birds existed. They are stable enough that evolution can count on them, not as instructions encoded in genes, but as parts of the environment that will reliably be present during development.

A small historical footnote adds human texture. Planetariums were built as public monuments to curiosity, places where city dwellers could see an artificial night sky even through urban haze. Emlen and others turned these domes into experimental instruments. The thought of a migratory bird inside a planetarium can feel almost surreal: a wild creature brought into a temple of astronomy to reveal how it perceives the heavens. Yet the lesson cuts both ways. The experiments did not merely show that birds use stars; they showed that animals can extract abstract structure from sensory experience, learning the *rules* of the sky rather than memorizing its details.

This is also where modern human life enters the picture. Light pollution erases stars and changes twilight. If a young bird grows up under sodium lamps and LED glare, the star field is dimmed and the polarization pattern can be distorted. The calibration cues may become noisier. It is too simple to claim that light pollution "breaks migration" in a direct line, but the calibration framework suggests a plausible vulnerability: if the reference system is degraded during sensitive periods, the magnetic compass may still function but may be less well aligned with geography. A migrant does not need perfect information to succeed; it needs enough consistency to keep errors from compounding.

On a crisp evening, away from city lights, the sky performs its slow, indifferent rotation, and twilight fades through a narrow band of colors that the human eye finds pretty and the bird's eye may find informative. The bird's magnetic compass is always there, an internal sensor that responds even when the stars vanish. But the sky provides the external guarantee that the compass is not drifting in meaning. The calibration is not a dramatic act; it is the quiet daily maintenance that turns a direction sense into a dependable reference, allowing a creature with no map in its beak to aim into darkness with the confidence of something that has, in its own way, checked its instruments.

## 7.3 Chasing the Drifting Pole

A compass feels like a promise: point it, and the world offers a stable answer. That promise is good enough for a weekend hike. It is less good for an animal that lives for decades, leaves home, and returns years later to a coast that has shifted under storms and sea-level change. The deepest complication is not the moving coastline. It is that the Earth's magnetic field—the very grid that makes a "magnetic signature" and a "magnetic compass" meaningful—is itself alive.

The field is generated by motion in the planet's outer core, a churning, conductive ocean of molten iron. From the surface, those motions are invisible, but their consequences are not. The position of the magnetic poles drifts. The strength of the field waxes and wanes. The angles of its lines tilt, subtly at some latitudes and dramatically at others. This slow evolution is called *secular variation*, and it takes place on the same timescale as many animal lives. For a loggerhead turtle that might not reproduce until its twenties or thirties, the field experienced at hatching can be measurably different from the field encountered at first nesting. For a seabird that migrates for fifteen years, the map changes during its career.

A human analogy helps, but only if we resist making it too neat. Imagine you learned to navigate a city as a child, storing a set of reliable cues: the river runs east-west; the church spire is a north-pointing marker; the train station is always down the hill. Then imagine that, while you were away, the river slowly changed course, the hill's slope subtly shifted, and the church spire rotated a few degrees every year. Returning with your childhood map would not cause disaster at once. It would cause small errors that compound: the street you expect to meet does not meet; the shortcut ends at a fence; the landmark sits just off where it should. Secular variation does that to magnetic navigation. It does not delete the map; it nudges it.

Humans who rely on magnetism have long known this. Navigators distinguish between *magnetic north* and *true north* and apply a correction called declination. Declination is not constant; it varies across the globe and changes with time. Aeronautical charts and nautical charts include declination information and often note the year so the correction can be updated. A ship captain who ignored this would not immediately sail into rocks, but over long distances the angular error would translate into miles.

The modern world solves this with models of the magnetic field. One of the most widely used is the World Magnetic Model (WMM), maintained for practical navigation and updated on a regular cadence. Its most recent full release, WMM2025, was issued on December 17, 2024 and is intended for use through late 2029. The scientific community also maintains the International Geomagnetic Reference Field (IGRF), a standard mathematical description of the main field and its slow change, used to compute field components anywhere on Earth across past epochs and near-future forecasts. These models are a reminder that the magnetic field is not background scenery; it is a dynamic geophysical variable, something people must periodically remeasure and re-estimate.

Animals, of course, do not download WMM updates. Yet many appear to navigate as if they had some method of keeping their internal system aligned with a moving target. The question is not whether secular variation exists, but what biological plasticity—the capacity to adjust based on experience—must exist for magnetic navigation to remain useful.

One source of plasticity is the simplest: do not treat magnetic information as absolute. A turtle that has imprinted on a "home signature" can treat it as a region rather than a point, searching broadly along a coast until other cues lock in. A salmon can use magnetic cues to reach the correct continental margin and then hand off to smell. A bird can calibrate magnetic headings against twilight cues and stars, effectively tying the magnetic compass to the Earth's rotational frame rather than to the capricious location of magnetic north. The field can drift, but the sky's rotation does not.

There is also a more subtle possibility: animals may update their magnetic map over time. 'Updating' can sound like a high-level cognitive act, but it can be as simple as repeated association. If an animal repeatedly experiences a particular magnetic environment in a place that is behaviorally important, the nervous system can retune what 'home-like' feels like. For a seabird that returns to the same breeding colony every year, annual revisits provide calibration opportunities. For a turtle that returns every few years, each nesting season could refresh the association between local cues and local magnetic conditions. In that sense, a long-lived animal might not be chasing the magnetic signature of its birth night exactly; it might be chasing a signature that has been periodically nudged toward the present.

Evidence that secular variation has biological consequences comes from precisely the kinds of population-level patterns that make the imprinting hypothesis testable. The logic is direct: if animals use magnetic signatures as targets, then changes in those signatures should shift where animals arrive, even if everything else about the landscape stays the same. Loggerhead nesting distributions along coasts have been analyzed in this spirit, linking long-term changes in nesting density to the convergence and divergence of magnetic isolines. Similar reasoning has been applied to salmon route choices: if juveniles imprint on magnetic conditions during an outward migration and adults later seek the same conditions, then the slow drift of those conditions could bias migration routes in predictable ways.

There is a crucial emotional point here, often lost in the technical description: an animal's "mistake" may not be an error at all. A turtle arriving at a beach slightly north of its natal site might be following its internal program faithfully, only to find that the field's contours have shifted. The animal is not wrong; the world is different. To the extent that magnetic cues are used, secular variation imposes a built-in uncertainty that no individual can eliminate. What evolution can do is build systems that are robust: systems that degrade gracefully rather than catastrophically.

Robustness often comes from redundancy. In practice, magnetic information is rarely the only guide. Many migrants use wave direction, current shear, temperature gradients, salinity fronts, odor landscapes, polarized light, star patterns, and learned topography. A magnetic map can be the large-scale scaffold while other cues provide fine-grained corrections. This is not hand-waving; it is a plausible computational strategy. In engineering, sensor fusion improves performance when each sensor has different failure modes. A magnetic sensor works under clouds and in darkness; a star compass works only when stars are visible; an odor cue works only when molecules arrive from a source. Combined, they can cover one another's gaps.

A historical anecdote from human navigation captures the same idea. During the age of sail, mariners did not trust a single instrument blindly. They combined dead reckoning, celestial navigation, soundings, and compasses, and they were wary of local magnetic anomalies that could deflect a needle. The practice of repeatedly checking and re-checking against independent references was not pedantry; it was survival. Birds that recalibrate their magnetic compass daily at twilight are practicing a comparable discipline, built into instinct and physiology rather than written in a manual.

The "drifting pole" itself is a vivid, almost theatrical expression of secular variation. The North Magnetic Pole has wandered across the high Arctic over the last century and, during some recent decades, has moved with striking speed. That movement is not a capricious darting of a point marker; it reflects deep processes in the core that shift the geometry of the field. For animals, the pole's motion matters less as a pin on a map than as a sign that the field's structure can change substantially within a lifetime, especially at high latitudes where gradients are steep. A small change in pole position can reshape local declination and alter the direction of "magnetic north" experienced at the surface.

If the pole drifts, what happens to an animal that uses a magnetic compass but calibrates it against geography? The answer depends on whether the animal's "north" is magnetic or geographic. Birds that calibrate using celestial rotation can, in effect, define their preferred headings in a geographic frame while still using magnetic sensory input to hold those headings in flight. That reduces the impact of changing declination, because the compass is regularly re-anchored to the sky's rotational axis. A bird does not need to know the word "declination" to correct for it. It only needs a habit of checking the magnetic compass against a cue that does not drift.

For animals using magnetic map cues (like intensity and inclination), the challenge is different. Map cues change not only in direction but in value. A turtle imprinted on a particular pair of values might later encounter that pair displaced along the coast. If the animal treats that pair as a strict target, it could be systematically drawn away from its natal site. If it treats it as one cue among many, the shift might only broaden the search or bias nesting slightly. Field evidence suggests the latter: distributions shift, but the animals do not abandon entire regions.

There is also a generational dimension. If imprinting occurs early and the field shifts over decades, then cohorts born in different years may imprint on slightly different magnetic "addresses" even at the same beach. That could produce a gentle spreading of return locations along a coast over time, or it could create a moving target that the population tracks. In this view, secular variation is not merely a problem to be solved; it becomes part of the ecology of a species, shaping patterns of dispersal, mixing, and genetic structure.

One might ask why evolution would ever 'choose' magnetic cues if they drift. The answer is that every cue drifts in some way. Stars are stable, but clouds block them. Odor signatures shift with rainfall, pollution, and river engineering. Coastlines change. Ocean currents shift with climate. The magnetic field is unusually dependable in one respect: it is available day and night, underwater and above, even in featureless environments. Its changes are typically slow enough that robust strategies can accommodate them, and that slowness can itself be exploited as a source of testable predictions. When researchers use secular variation as a natural perturbation, they are doing something like what physiologists do when they change temperature to reveal enzyme dynamics: using a controlled, real-world change to expose underlying mechanisms.

A final layer of plasticity may occur not in the "map" or the "compass" separately, but in how the brain arbitrates between them. Cue-conflict experiments in birds and other animals show that the preferred cue can change depending on context: time of day, weather, age, migratory stage, and the reliability of information. A nervous system that can assign weights—trust magnetism more when the sky is overcast; trust stars more when the horizon is visible; trust smell more near shore—is better suited to a changing planet. That kind of weighting can be innate, learned, or both. It can also be updated: if a cue repeatedly misleads, the animal can downweight it. Even without conscious deliberation, that is a form of experience-dependent calibration.

The practical consequences are surprisingly immediate. Conservationists relocating hatchling turtles away from bright city lights must consider what magnetic environment those hatchlings experience during their first crawl and swim. Hatcheries moving eggs a few kilometers along a coast may inadvertently shift the magnetic signature imprinted during a sensitive window. Fish raised in hatcheries and released into rivers may imprint on magnetic conditions at release sites rather than natal gravels, potentially altering return patterns. None of these outcomes is inevitable, but the possibility arises precisely because the magnetic field is both informative and variable.

The drifting pole, then, is less a curiosity of geophysics than a pressure test of biology. Any animal using magnetism faces a choice between rigidity and flexibility. Rigidity gives crisp, efficient behavior in a stable world; flexibility gives resilience in a changing one. The Earth offers no stable world. It offers a slow restlessness in its core, a field that is always becoming something slightly else.

On a winter night in the high Arctic, the aurora can arc across the sky in curtains, an electric show driven by interactions between solar wind and the magnetosphere. It is a reminder that magnetism is not merely a line drawn on a map; it is an active feature of the planet's relationship with space. Under that same sky, far south, a migrating bird departs after twilight, holding a steady direction while the magnetic field beneath it shifts by fractions of a degree from year to year. The bird's success depends not on a perfect, unchanging compass, but on a system that can keep working as the reference itself slides—a living solution to a moving target.

## Chapter 8

# A World Full of Noise

*Imagine trying to read a compass while standing directly on top of a massive loudspeaker; this is the sensory reality for many animals navigating the Anthropocene. We investigate how modern 'electrosmog' and our sleepless, illuminated cities are systematically jamming the delicate quantum sensors of migratory species, turning their evolutionary guidance systems into sources of fatal confusion.*

## 8.1 The Static in the Signal

A migrating bird does not carry a compass in the familiar sense: no needle, no dial, no tiny map tucked under a feather. Yet on a clear night in autumn, a European robin can step into a cage in a laboratory, hop with unmistakable restlessness in one preferred direction, and keep doing so night after night as if some internal pointer has been set. What makes that feat unsettling—in the best scientific way—is that the relevant physical signal is extraordinarily weak. Earth's magnetic field is about 50 $\mu$T in much of Europe, roughly the strength you would measure a few centimeters away from a fridge magnet only after you have moved far enough for the magnet to feel almost gone. A robin reads not just the presence of that field but its geometry: the tilt of the field lines, the inclination that changes with latitude. For a long time this was treated as an exotic sense, a black box that must be robust because it works so reliably in nature. Then a peculiar vulnerability began to show itself. The compass could be knocked off course, not by strong magnets, but by something closer to background hiss.

The first clue came from an experimental trick that sounds too modest to matter: adding a very weak oscillating magnetic field—a tiny wiggle on top of Earth's steady field. In many species of nocturnal migratory birds, this extra field can be so small that it is hard to believe it has any biological relevance at all, and yet it can scramble the birds' orientation. The striking part is not merely that the compass can be disrupted; it is *how* it can be disrupted. The effect depends on frequency. That single detail opens the door to a mechanism that would once have seemed wildly implausible in warm, wet tissue: quantum spin dynamics.

To make sense of that claim without mysticism, it helps to start with an image of a chemical reaction that comes with a built-in yes/no switch. Certain light-sensitive molecules can be excited by a photon so that an electron is transferred within the molecule, leaving behind two fragments that each carry an unpaired electron. Chemists call this a *radical pair*. Unpaired electrons behave like tiny bar magnets, not because they are little spinning balls, but because they possess a quantum property called *spin*. Spin is quantized: it can point in certain allowed ways relative to a field. In a radical pair, the two electron spins can be arranged in a *singlet* state (paired, opposite in a way that makes the total spin zero) or a *triplet* state (aligned in a way that gives the pair a net spin). The singlet and triplet states are not merely labels; they can lead the chemistry down different pathways. If the pair is in the singlet state at the moment the reaction "decides," it may recombine into one product; if it is in the triplet state, it may separate or form another product. The proportion of singlet versus triplet products is a kind of chemical readout.

Now add Earth's magnetic field. A magnetic field changes the energy landscape of spins and nudges the radical pair to interconvert between singlet and triplet character as time passes. Importantly, the nudging is not the same for all molecular orientations. If the molecule is oriented one way relative to the field, the spin dynamics unfold slightly differently than if it is rotated. That makes it possible, at least in principle, for a biological system to extract directional information: a pattern of chemical yields across many molecules with different orientations could form a compass-like signal.

The idea sounds delicate because it is. The relevant energies are tiny; thermal agitation is relentless; cells are full of magnetic nuclei and fluctuating fields. So when behaviourists found that weak, oscillating magnetic fields at certain radio frequencies could disrupt birds' magnetic orientation, the result looked less like an oddity and more like a signature. If the compass depends on radical-pair spin dynamics, then adding an oscillating field at the right frequency can drive transitions between spin states—rather like pushing a child on a swing, but only if your timing matches the swing's natural rhythm. In physics this is the logic of resonance: an external oscillation can have an outsized effect when it matches the system's characteristic frequency.

A useful way to picture the resonance is to imagine the radical pair as a microscopic metronome that ticks in quantum time. The singlet and triplet states are separated by an energy gap that depends on magnetic interactions within the molecule and on the external magnetic field. If an oscillating magnetic field supplies energy at a matching frequency, it can promote mixing between these states, blurring the directional differences that the bird is trying to detect. A narrow, carefully tuned frequency can do this; so can broadband noise that contains many frequencies at once. The second possibility matters because it turns a laboratory manipulation into an ecological threat.

That ecological threat was sharpened dramatically by a set of experiments from Henrik Mouritsen's group at the University of Oldenburg, reported by S. Engels and colleagues in 2014. The headline, stripped of all caveats, sounds almost like a fable about modernity: robins in a city could not find north until the researchers built them a Faraday cage. The reality is both more precise and more unsettling.

The birds were European robins, tested in the standard way used for decades: each bird was placed in a circular orientation cage (an *Emlen funnel*) lined with scratch paper. A migratory bird with directional restlessness hops in the direction it wants to travel, leaving a pattern of scratches that can be scored. The tests were carried out in wooden huts near the university campus. The huts were not inside a power station or under a radio mast; they were simply in an environment saturated with the ordinary electromagnetic emissions of a modern campus: computers, networking equipment, switching power supplies, lighting systems, security devices, and the invisible chatter of electronics that few humans ever notice.

In those huts, the robins failed to show their usual magnetic orientation. They became directionally confused, as if the compass had been removed. The researchers then did something ingenious and, in hindsight, almost obvious: they converted one hut into a shielded room by lining it with grounded aluminium screening, essentially creating a Faraday cage that attenuates time-varying electromagnetic fields. The effect was immediate and dramatic. In the shielded, grounded hut, the robins regained their proper migratory orientation. When the screening was present but not grounded, the improvement was lost. When the hut was grounded and quiet but the researchers deliberately introduced broadband electromagnetic noise, the confusion returned. They also tested in a rural location with much lower background electromagnetic noise; there the birds oriented normally without any shielding. The pattern was not a vague correlation but a controlled demonstration: under urban electromagnetic conditions, the compass could be blinded; remove the noise, and it worked again.

Several details matter for understanding why this is so provocative. The disruptive background noise was in the approximate range from tens of kilohertz to a few megahertz—a band that sits awkwardly between the frequencies most people associate with "radio" and the gigahertz signals of mobile phones and Wi-Fi. The fields were weak, far below levels that would heat tissue. This is not the world of microwave ovens or burns. It is the world of subtle oscillations that, to a radical pair, can be the right kind of push. The results also implied that the noise need not be a single pure tone. The birds were affected by broadband noise: a smeared-out spectrum like the hiss between stations. That makes the problem harder to avoid, because broadband noise is exactly what many electronic devices generate as a by-product of switching currents on and off.

At this point the story divides into two parallel dramas: one in biology and one in engineering. In biology, the question is whether the avian magnetic compass truly relies on radical pairs, most plausibly involving cryptochromes—light-sensitive proteins found in the retina—and whether such a mechanism can be sensitive enough, in vivo, to be disrupted by such small oscillating fields. In engineering, the question is how much of the electromagnetic environment we have inadvertently filled with precisely the kind of noise that would be maximally inconvenient for a quantum compass.

The biological drama is sometimes misunderstood because the word "quantum" tends to attract exaggeration. Nothing here requires long-lived quantum coherence of the kind discussed in quantum computers. The radical-pair mechanism is quantum in a simpler, unavoidable way: spin states are quantum states. The interconversion between singlet and triplet character is governed by quantum rules even if it occurs rapidly and in a noisy environment. What matters is not that the system is isolated from noise; it is that the relevant spin dynamics are fast enough and structured enough that weak magnetic perturbations can bias the chemical outcomes by a detectable amount. That is a demanding requirement, but it is not obviously impossible. In 2021, for example, Jingjing Xu, Lauren Jarocha and colleagues reported magnetic sensitivity in purified cryptochrome 4 from the European robin under laboratory conditions, showing that the protein's photochemistry can, in principle, respond to Earth-strength fields. The chain from a photochemical yield difference to a perceptual experience in a living bird remains a frontier, but the plausibility of the core physics is now far less speculative than it once seemed.

The engineering drama is, in some ways, the more immediate. Humans have developed elaborate regulatory frameworks for electromagnetic exposure, but these are overwhelmingly designed to protect humans from established harms, especially tissue heating at higher frequencies. A magnetoreceptive animal faces a different kind of risk: not damage, but deception. A compass that is not broken but misled is a dangerous tool. A bird does not need to be injured to be lost.

To appreciate the scale of the mismatch, consider the difference between a field that is biologically *safe* and a field that is biologically *informative*. A typical human safety guideline might tolerate a certain power density because it does not measurably raise body temperature. But a radical-pair compass is not a thermometer. It is closer to a radio receiver that is trying to pick out a faint station. You can flood the airwaves with static without burning the receiver. You simply prevent it from doing what it evolved to do.

A concrete analogy helps. Imagine trying to navigate a city using a cheap handheld compass. In the countryside it works well enough. In the city, you may find it behaving oddly near steel structures, power cables, or even a car. Now imagine that your compass needle is not a piece of magnetized metal but a chemical reaction whose "needle" is the balance between singlet and triplet products. It is as if your compass is not pulled by the field directly but listens to a faint rhythm set by the field. A broadband radiofrequency field is then like someone playing a drum kit in the same room, throwing in beats at many tempos. Even if none of those beats is loud in an absolute sense, their presence at the wrong moments can overwhelm the timing cues that matter.

The resonance idea can be made more intuitive with one additional piece of physics. In an external magnetic field $B$, an electron spin tends to precess at a characteristic angular frequency $\omega = \gamma B$, where $\gamma$ is the gyromagnetic ratio. For Earth-strength fields, this frequency lies in the megahertz range. That happens to overlap with the frequency band where the Oldenburg team measured strong attenuation effects with shielding and where disruptive noise was implicated. A narrowband field at the Larmor frequency can be a particularly efficient way to perturb spin dynamics, but broadband noise that includes that region can still do damage. If the animal's compass relies on detecting subtle differences in reaction yields as a function of direction, any additional mixing of singlet and triplet states can flatten those differences. The compass does not point the wrong way in a stable manner; it becomes noisy. A needle that jitters is worse than a needle that is consistently off by five degrees.

The most troubling aspect is that the bird cannot easily "know" its magnetic sense is being corrupted. When humans lose GPS signal, the phone often tells us. When a robin's compass is degraded, the rest of its sensory world may look perfectly normal. The stars may be hidden by cloud; landmarks may be unfamiliar; odours may be unhelpful over open water. In that context, a degraded compass is not merely an inconvenience. It is a risk multiplier. A small error repeated over hours becomes a large displacement. A hesitation at the wrong time can mean arriving late to a stopover site when food is scarce, or being forced down in poor habitat.

The Oldenburg result also carries a psychological sting for those of us who live surrounded by devices: the disruptive noise was not produced by some dramatic piece of infrastructure. It was "background." The modern electromagnetic environment is not a single transmitter you can point to and regulate; it is an emergent property of countless gadgets, cables, converters, and systems that leak energy as a side effect. The ecological footprint is therefore distributed. It is not always visible in the way a tower or a wind turbine is visible. It may be closer to a glow that spreads from every building.

There is an additional twist that makes the problem harder to dismiss. If the magnetic compass depends on radical pairs, then the vulnerability is not a pathological quirk of one species. It is, in principle, shared by any animal that uses a similar photochemical mechanism. That includes not only birds but potentially other vertebrates and insects. The same logic that makes the radical-pair mechanism attractive—it can, in theory, provide a directional sense without requiring magnetic material—also makes its failure mode universal: oscillating fields at the wrong frequencies can interfere with spin dynamics anywhere.

At the same time, it would be dishonest to present the story as closed. Field measurements of electromagnetic noise are complex. Urban environments vary enormously. Some sources are intermittent. The coupling between external fields and the relevant molecules inside an animal is not straightforward. Even in the lab, behavioural responses depend on light conditions, species, and experimental details. The appropriate lesson is not that every city is an automatic navigational death trap, but that a sensory channel we once assumed to be stable can be fragile in a specifically modern way. The fragility has a mechanistic explanation grounded in physics, and it has been demonstrated under controlled conditions in a way that demands attention.

One can feel the poignancy most sharply by returning to the robin in its funnel. The animal is not performing an abstract task. It is acting out an inherited program that, for countless generations, has worked well enough to keep populations connected across continents. The bird's restlessness is a behavioural trace of an internal map of seasons: leave now, fly this way, avoid that. In the quiet of a rural night, the compass emerges from chemistry as if it were a sense organ like vision, except that its stimulus is a planetary field that never switches off. In the electromagnetic bustle of a campus, the compass can vanish, not because the bird has changed, but because the medium has. A world that is loud in radio becomes, for a radical-pair compass, a world where north is washed out.

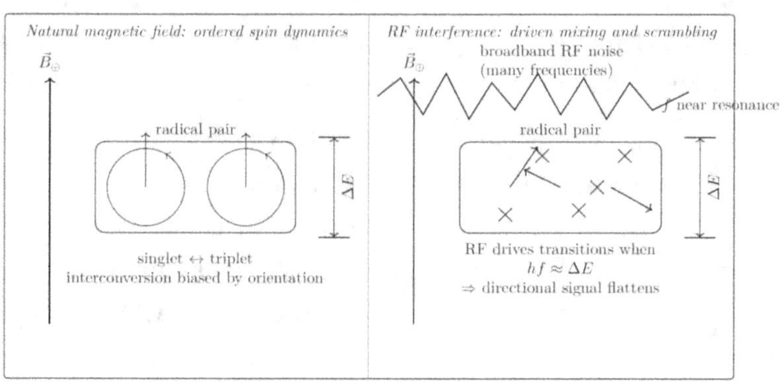


The unsettling implication is not that the magnetic compass is feeble, but that it is *selective*. A selective sense can be exquisitely informative in the environment it evolved in, and exquisitely vulnerable in an environment that jams its particular channel. The robin's compass does not fail because the city is stronger than nature; it fails because the city is noisy in exactly the way a spin-based sensor would least tolerate. In a landscape where the electromagnetic spectrum has become part of the habitat, silence is no longer just the absence of sound. It is the absence of static in the signal.

## 8.2 Fractured Twilights

Twilight is not simply the day turning off. For many migrating birds it is a daily calibration ritual, a brief window when the sky offers a set of reliable geometric cues—the sun's position slipping below the horizon, the changing gradient of brightness, and a pattern of polarization that sweeps across the dome overhead. In that window, a young bird can align its internal reference frame: which way is "toward the setting sun," how that relates to north and south, and how to reconcile what it sees with what it senses magnetically. A few minutes later the sun is gone, stars emerge, and the bird takes off into darkness with its instruments freshly set.

Modern nights have begun to tinker with that ritual. City light does not merely brighten the ground; it spills upward and outward, rebuilding twilight after it should have ended. The horizon no longer darkens cleanly. The sharp line between the sunlit atmosphere and the night sky becomes blurred by domes of skyglow. Searchlights and illuminated towers carve bright wedges through cloud. Even on clear nights, the spectrum of the sky is no longer the spectrum that shaped the sensory biology of nocturnal migrants.

This matters for magnetoreception because the avian magnetic compass is not, in the leading view, a standalone device. It is a light-dependent sense. The same photons that let a bird see can also trigger the chemistry that lets it *feel* magnetic direction. That dependency creates a vulnerability that is both elegant and cruel: change the night's light, and you may change the compass itself.

The best way to keep the story grounded is to follow a bird through a single evening, and to separate three roles that light plays. First, light provides ordinary visual information: silhouettes of trees, reflections off water, the glow of a city. Second, twilight light provides *celestial calibration cues*: the sun's azimuth at sunset and the polarization pattern near the horizon, which birds can use to set a directional reference. Third, light acts as a biochemical *switch* that turns on (or tunes) the radical-pair chemistry thought to underlie the magnetic compass, probably in cryptochrome proteins in the retina. Artificial light at night can interfere with all three roles at once, which is why it is so hard to disentangle distraction from genuine sensory jamming.

Begin with the calibration cues, because they anchor the drama in something any human has felt: twilight has a character. On a countryside evening, the western horizon holds its brightness for a while, then drains away; the east deepens into blue; the first bright stars puncture the sky; the rest follow. Near sunset and sunrise, sunlight scattered through the atmosphere becomes strongly polarized. Polarization is a property of light waves: the electric field oscillates preferentially in one plane rather than in all directions. Humans are mostly blind to polarization, but many animals are not. For a bird with polarization sensitivity, the twilight sky contains a structured pattern that is tied to the sun's position even when the sun itself is hidden.

Cue-conflict experiments over decades—often done with migratory songbirds in orientation cages—suggest that birds do not treat magnetic direction, sunset direction, and star direction as independent compasses that always agree. Instead they appear to use one cue to calibrate another. Twilight, especially the band of sky near the horizon, is repeatedly implicated as a crucial setting event. Remove the horizon view or scramble the polarization pattern and the bird's later magnetic orientation can shift or become unreliable, as if the reference frame has been mis-set. That is a subtle point: the bird can still *sense* magnetic direction, but what it calls "north" may have been recalibrated against a distorted sky.

Now consider what skyglow does to the horizon. In a brightly lit urban area, the horizon may be permanently luminous, with gradients dominated not by the sun's last light but by the geometry of street-lights and buildings. The polarization pattern, too, can be weakened because artificial light is often less polarized than scattered sunlight, and because haze and aerosols in urban air can change scattering conditions. The bird still experiences a twilight, but it is a fractured twilight: the cues are present, but altered in intensity, geometry, and reliability. Calibration becomes a negotiation with a sky that lies.

The second role of light is biochemical, and it loops back to the radical-pair story without repeating it. The key idea is that cryptochromes are photoreceptors: they undergo chemical changes when they absorb light, particularly in the blue to green range. In the radical-pair hypothesis, those light-driven chemical changes create radical pairs whose spin dynamics depend on the magnetic field. If the illumination is too dim, perhaps too few molecules are activated. If it is too bright, the system may saturate or run into alternative reaction pathways. If the wavelength is wrong, the relevant photochemistry may not be triggered at all. Light, in other words, is not just an external cue; it is part of the compass's operating conditions.

That dependence makes artificial lighting unusually potent. A streetlamp is not a miniature sun. Its spectrum is shaped by technology. Older low-pressure sodium lamps emitted a narrow band of yellow light, which is visually harsh but spectrally simple. Many modern white LEDs are built from a blue LED plus phosphors that convert some of the blue to longer wavelengths, leaving a characteristic spike in the blue and a broad hump across green to red. To a human eye, both can look like "light." To a photopigment, the difference is decisive.

Cryptochromes tend to absorb strongly in the blue. That does not prove that blue-rich light is uniquely harmful, but it makes the overlap plausible: the very wavelengths that many LEDs emit efficiently are the wavelengths that most effectively drive cryptochrome photochemistry. A bird flying over a city bathed in blue-rich light may therefore be carrying a magnetic compass that is being forced into a different operating regime. Instead of being activated by the dim, spectrally filtered light of natural night and twilight, it is being activated by a bright, unnatural spectrum from below.

This is where many popular accounts become too simple, suggesting that artificial light merely "distracts" birds. Distraction is real. Anyone who has watched moths or seabirds spiralling around a bright lamp has seen the pull of light. But magnetoreception adds another layer: a bird can be both attracted visually and, at the same time, deprived of a reliable magnetic bearing. In an urban glow, a bird might not simply choose the wrong direction; it might be prevented from choosing any direction confidently.

A concrete, almost cinematic example comes from New York City's "Tribute in Light," the annual memorial installation that projects two intense vertical beams into the night sky. In 2017, researchers using radar and acoustic monitoring documented that when the beams were on, large numbers of migrating birds became concentrated and disoriented around the light, with behaviors consistent with circling and prolonged flight. When the lights were temporarily switched off, the disruption diminished rapidly and the birds dispersed, resuming more typical migration patterns. The power of the example lies not only in its scale but in its reversibility: change the light, change the behavior, within minutes. It is a reminder that nocturnal migration is an active, responsive process. Birds are not inert projectiles crossing the continent; they are constantly sampling cues and updating decisions.

Such dramatic installations are unusual, but the same principle applies in a quieter form across thousands of buildings. Many birds migrate at night and are drawn toward bright areas on broad geographic scales, a pattern supported by analyses of weather radar data such as the study by James McLaren, Jeffrey Buler and colleagues (2018) in the northeastern United States. Intriguingly, those analyses also suggest a more complicated local picture: while migrants may be more abundant in or near brighter regions at large scales, densities can drop within a few kilometres of intense light sources, consistent with local avoidance, disruption, or poor stopover habitat. The point is not to treat light as a simple magnet. Light reshapes the landscape of decisions at multiple scales: where to fly, where to land, whether to keep going, whether the night feels safe.

Magnetoreception threads through those decisions because it supplies the directional backbone when other cues are degraded. On a clear rural night, stars and magnetic cues can complement each other; on cloudy nights, magnetic cues become more valuable. Urban skyglow often arrives with haze and low clouds that reflect light back down, erasing stars and turning the sky into a luminous ceiling. That combination is especially treacherous: celestial cues are removed, visual landmarks are distorted, and the magnetic compass may be operating in a photochemical regime it did not evolve for. Many of the notorious mass-collision events at illuminated buildings occur on overcast nights for precisely this reason: the birds are funnelled into the bright zone, circle, exhaust themselves, and strike glass or structures.

There is a further layer, less visible but biologically profound. Night is not just a time for navigation; it is a physiological signal. Hormones such as melatonin rise at night in many animals, conveying information about darkness and time. Exposure to light at night can suppress melatonin, disrupting circadian rhythms and sleep. In 1994, work on migratory birds showed that light at night significantly diminished nocturnal melatonin levels; intriguingly, altered magnetic field conditions were also reported to affect melatonin in that study, hinting that magnetic and light cues might intersect not only in navigation but also in physiology. Even without leaning heavily on that contested intersection, the broader message is solid: artificial light can make night biologically less like night. For a migratory bird that must maintain nightly endurance and daytime recovery over weeks, chronic disruption of night physiology can translate into poorer timing, reduced fattening, and diminished performance. A bird does not need to be "lost" to be harmed; it can simply be worn down.

One might ask whether birds can adapt by ignoring magnetic cues and relying more on other senses in lit environments. Sometimes they can. Migrants have multiple compasses and multiple sources of information: stars, the sun, polarized light, landmarks, odours, wind. But switching cues is not cost-free. Many of these cues require clear skies, experience, or proximity to known terrain. A first-year bird crossing an unfamiliar region at night cannot conjure a memory-based map. In fog, stars disappear. In a featureless landscape or over water, landmarks vanish. Wind can push birds off course; without a compass, compensating becomes guesswork. The magnetic sense is not a luxury add-on. It is often the most reliable cue in the very conditions—darkness, cloud, long-distance travel—that define nocturnal migration.

Artificial light can also create a particularly cruel feedback loop at dusk. If the horizon is brightened and distorted, calibration becomes uncertain. If the bird then departs into a sky where stars are hidden by reflected city light and clouds, it has fewer independent checks on its bearing. If the magnetic compass is itself perturbed by the spectral conditions, the bird is not merely missing a cue; it is working with corrupted information. The result is not necessarily a dramatic wrong-way flight that would be obvious to an observer. It can be something more insidious: increased scatter in headings, hesitation, repeated course corrections, extra kilometres, delayed arrival, and elevated mortality risk spread across a population rather than concentrated in a single spectacle.

The emotional sting for humans is that many of the sources are signs of prosperity and safety. Streetlights are installed to prevent accidents and crime. Brightly lit buildings signal activity and welcome. LEDs are celebrated for energy efficiency. None of this is malicious. The conflict arises because birds experience light not as decoration but as environment, and because their sensory systems are tuned to a night that used to be darker, redder, and more predictable.

It is tempting to look for a single villain wavelength—to declare blue light uniquely responsible and treat redder lighting as a cure-all. Reality is messier. Spectrum matters, but so do intensity, directionality, timing, and context. A poorly shielded warm lamp that pours light upward can still create skyglow and erase stars. A well-designed white light aimed downward and dimmed late at night may reduce harm more than a poorly designed red light that is simply brighter. Some birds may be more sensitive than others. The magnetic compass itself likely has an intensity window: too little light and it cannot operate, too much and it may saturate. Natural nights already vary with moon phase and cloud cover; artificial light adds a new baseline and new extremes.

What remains consistently actionable is the idea that the sky is part of habitat. A forest can be intact on the ground and still be navigationally hostile overhead. The sky can be made less legible by light that spills upward, and the internal chemistry of the compass can be pushed into unfamiliar territory by spectra that over-activate its photoreceptors. When those pressures combine during migration seasons, the result is a nightly toll that is easy to miss precisely because it is spread across time and space.

The most haunting image is not a single bird colliding with glass, though that is tragedy enough. It is the altered twilight itself: a horizon that no longer darkens properly, a polarization pattern diluted by glow, a night sky without stars, and a retina flooded with the wrong photons. In that altered light, a magnetic sense that evolved to be partnered with dusk becomes something like a finely tuned instrument played in a room full of glare, still capable of music but constantly at risk of slipping out of tune.

## 8.3 Restoring the Quiet

Quiet is usually imagined as the absence of sound: a forest after snowfall, a shoreline when the wind drops, the sudden hush when a city loses power. For magnetically sensitive animals, quiet also has an electromagnetic meaning. It is the difference between a night where the sky is dark enough to keep its geometry, and a night where skyglow erases the horizon's cues. It is the difference between a magnetic compass that can extract directional information from delicate spin chemistry, and one that is smeared into indecision by the broadband chatter of electronics. The uncomfortable part is that the electromagnetic environment is now a human-made habitat feature, as real as a road or a fence, but far easier to overlook because it is mostly invisible to us.

Conservation has always been about what animals can perceive and use. Early bird protection laws were sparked by something tangible and brutal: the millinery trade that decorated hats with feathers. The harm was visible in the bodies of birds and in emptied colonies. The response was cultural and legal: change what people buy and what hunters can sell. The electromagnetic challenge is different. The harm is not a single act of killing, but the steady accumulation of small interferences: light that leaks upward every night, radiofrequency noise that seeps from a web of devices. The response, too, is different. It asks for design rather than prohibition, for timing rather than permanence, and for a new habit of thinking: treat the night sky and the electromagnetic spectrum as shared resources.

A useful starting point is to abandon the fantasy of a perfect fix. There is no realistic way to return continents to pre-electrical darkness, and there are obvious human needs for lighting and communication. The more practical question is where quiet matters most, and how to supply it without sacrificing safety or modern life. Fortunately, migration biology gives clear leverage points. Birds do not move randomly across the landscape. They funnel along coastlines and river valleys, concentrate over passes, and cluster in cities and parks as stopover habitats. Their heaviest movements happen on particular nights, shaped by weather and season. This makes targeted mitigation possible: not an all-year blackout, but intelligent restraint at the times and places when it buys the most biological benefit.

The most persuasive evidence for this kind of targeted action comes from the same dramatic city-scale light sources that make the problem so obvious. New York City's Tribute in Light has become an accidental experiment in migration management. When those beams are switched on during peak migration, large numbers of nocturnal migrants can become trapped in the illuminated zone, circling and calling. When the lights are switched off, even briefly, the concentration breaks and the birds disperse. The crucial lesson is not merely that bright light attracts birds. It is that turning down the light can produce an immediate, measurable improvement. This is conservation at the timescale of minutes.

That immediacy is what makes the "Lights Out" idea so attractive. The basic proposition is simple: during migration seasons, turn off or dim non-essential lights at night, particularly in tall buildings where illuminated interiors and decorative exterior lighting pull birds into collision risk. The program's origins are civic rather than academic; many versions were promoted by local Audubon organizations and partner groups long before the details of cryptochrome chemistry were widely discussed. The focus was practical and humane: fewer dead birds under windows in the morning. As the science of light pollution and navigation has matured, the same simple intervention has gained a deeper rationale. Turning off lights does not just reduce collisions; it restores information to the night: stars become visible, the horizon regains contrast, and the photochemical operating conditions of a light-dependent magnetic compass become less extreme.

Modern versions of Lights Out are increasingly guided by forecasting rather than fixed calendars. Bird migration is not uniform; it comes in pulses. In North America, projects such as BirdCast use weather radar and atmospheric data to estimate when and where large movements of nocturnal migrants are likely to occur, and to communicate "high migration" nights to cities and building managers. That matters because it translates an ecological event into an operational decision. A building manager does not need to commit to months of inconvenience. They can respond to a handful of nights when the benefit is greatest. In practice, this often means turning off unnecessary exterior lighting and dimming interior lights in unoccupied spaces, typically during the late-night hours when migratory flights are heaviest.

A second idea, closely related but more engineering-driven, is to redesign outdoor lighting so that it illuminates what needs illumination and spares what does not. The International Dark-Sky movement and lighting engineers have converged on a set of principles that are almost embarrassingly commonsensical once stated: use light only where it is useful, aim it at the target rather than the sky, keep intensity as low as practical, control it with timers or motion sensors, and choose warmer spectra when possible. These principles are often presented as a package of co-benefits: lower energy bills, safer streets with less glare, better sleep for humans, and less ecological disruption. The migratory-bird case adds urgency because it links these design choices to life-and-death navigation.

The details matter. Poorly designed lighting can increase glare and decrease visibility for drivers and pedestrians, producing the paradox of brighter lights making streets less safe. A fully shielded fixture that directs light downward can allow lower wattage for the same useful illumination, while also reducing skyglow. Motion sensors can provide bright light only when people are present, rather than all night. "Warm" lighting, with less blue content, is often recommended because blue light scatters efficiently in the atmosphere (contributing disproportionately to skyglow) and because many biological photoreceptors, including those involved in circadian regulation, are especially sensitive to blue. For magnetoreception, the spectrum issue may be even more direct: cryptochromes absorb strongly in the blue, so reducing blue-rich emissions can plausibly reduce the risk of overstimulating the very photochemistry that underpins the magnetic compass. The point is not that red light is magically safe, but that spectrum is an adjustable parameter, and adjusting it can shift the night back toward something biologically legible.

The quiet that matters for birds is not necessarily complete darkness. Many species migrate under moonlight and navigate perfectly well. What matters is avoiding extreme, poorly controlled, upward-directed light that destroys contrast cues and creates traps. There is a difference between a lit path and a luminous sky. In much of modern lighting, the sky is the unintended casualty.

A third idea is spatial: identify and protect *dark corridors* and *dark refuges*. Migratory birds concentrate along flyways and rely on stopover habitats where they rest and refuel. A wetland or a patch of woodland inside an urban region can be ecologically rich but navigationally compromised if it is surrounded by intense light. Treating the electromagnetic spectrum as habitat means asking not only "is the wetland intact?" but "can birds find it, and can they land safely?" City planners and park managers can use lighting ordinances and fixture choices to keep key habitats darker, particularly during migration windows. In practical terms, this might mean reducing lighting along shorelines, dimming sports fields when not in use, shielding lights near water, and avoiding floodlighting that spills into the sky.

Radiofrequency electromagnetic noise is harder to manage because it is less visible and less centralized than light, but the principles of habitat thinking still apply. The Oldenburg experiments with European robins showed that shielding and grounding could restore magnetic orientation in an environment where background electromagnetic noise otherwise disrupted it. That does not mean cities should be wrapped in aluminium mesh. It does suggest that electromagnetic quiet can be engineered locally, and that the electromagnetic environment can be measured, mapped, and managed like any other pollutant. The analogy is not to banning electricity, but to the evolution of noise ordinances and building codes: once society recognized that chronic sound affects health and quality of life, it began to engineer quieter machines, better insulation, and zoning rules.

In the electromagnetic realm, the equivalent is to reduce unnecessary emissions at the source. Many devices generate broadband electromagnetic noise as a by-product of switching currents rapidly. Better filtering and shielding in power supplies, improved cable design, and attention to grounding can reduce radiated and conducted emissions. Industry already does some of this for electromagnetic compatibility: electronics must not interfere excessively with other electronics. The ecological twist is to treat wildlife sensory systems as part of the compatibility landscape. A bird's magnetic compass does not file a complaint with a telecommunications regulator, but it can still be jammed.

One can imagine, in the near term, a set of "quiet zones" that are modest in scope but strategically placed: reserves, coastal headlands, known stopover sites, and migratory bottlenecks where the benefits of reduced electromagnetic noise might be greatest. These would not necessarily require draconian restrictions. They could begin with measurement campaigns to establish baseline radiofrequency noise levels, followed by targeted interventions: moving particularly noisy equipment, improving shielding on infrastructure near sensitive sites, or designing new facilities with wildlife in mind. The most realistic early wins are likely to be small and local, but local changes can matter if they occur at ecological pinch points.

There is also an ethical dimension that is easy to miss. Many mitigation steps are framed as sacrifices: turn off your lights, dim your building, accept a darker street. In practice, much urban lighting is waste. Lights shine into empty offices at 2 a.m. because no one has bothered to install controls. Decorative uplighting is used by default rather than necessity. Glare-heavy fixtures waste light by sending it sideways and upward. Reducing these is not a loss of function; it is a recovery of purpose. The same is true for electromagnetic noise. Many emissions are not the price of communication; they are leakage. Quiet is often an efficiency gain.

A historical anecdote makes this shift in values vivid. When cities first electrified, brightness was a symbol of progress. Early photographs of illuminated streets celebrated modernity: the night conquered, the city made safer and more active. Only later did people begin to notice what had been lost: the Milky Way erased, insects swarming lamps, birds colliding with towers, sleep patterns altered. The first generation experienced light as liberation; later generations began to experience it as clutter. The arc is familiar from other environmental stories. Coal smoke was once the smell of industry; later it became a health crisis. Leaded petrol was once a technological triumph; later it became a neurotoxin in children's blood. In each case, the solution was not nostalgia but redesign.

A pragmatic conservation strategy therefore looks less like abstinence and more like craftsmanship. It asks engineers, architects, city officials, and citizens to treat light and electromagnetic emissions as materials with consequences. It asks for controls, shielding, and spectra in the same way that water management asks for filtration and runoff control. It also asks for timing. Migration is seasonal. A city that dims a handful of nights in spring and autumn, guided by forecasts, can provide a disproportionate benefit. A building that turns off decorative lighting after midnight can reduce risk while remaining brilliantly lit when people are awake to see it.

An emerging feature of this new approach is public participation. People can join Lights Out campaigns by drawing blinds, turning off porch lights, and advocating for better streetlight design. Communities can celebrate darker nights as civic events: star parties, night walks, and the simple pleasure of seeing constellations. In a subtle way, this is conservation that feeds back into human culture. When people can see the night sky again, they acquire an intuitive sense that darkness is not emptiness but atmosphere, not absence but structure.

The question of radiofrequency quiet is more technically demanding, but the same cultural shift can occur. Once people learn that some animals use the magnetic field through light-driven chemistry, the electromagnetic environment stops being an abstraction. A router, a switching power supply, a poorly shielded cable becomes part of an ecology. The point is not to treat everyday devices as villains but to recognize that design choices accumulate. A city can be made electromagnetically cleaner in the same incremental way it can be made quieter and less polluted: standards, incentives, and attention to details that used to be ignored.

None of this guarantees a perfectly navigable world for every migrating animal. Weather will still kill birds. Glass will still be dangerous. Habitat loss will still matter. Yet the electromagnetic approach has a rare quality in conservation: it often offers immediate benefits without demanding vast land purchases or long timelines. When the lights go out, birds stop circling. When glare is reduced, the sky darkens. When fixtures are shielded, the stars return. The night becomes more interpretable.

Quiet, in this sense, is not a romantic indulgence. It is an engineered condition: a night sky with contrast, a horizon that behaves like a horizon, a spectrum that does not overwhelm sensitive photoreceptors, and an electromagnetic background that does not drown out a planet-sized signal. It is the humility of making room for senses other than our own, and the practical wisdom of realizing that much of what we broadcast into the dark is not necessary for us, but may be decisive for everything that navigates by night.

Chapter 9

# Beyond Biology: Robots and Resiliency

*For decades, the idea that animals could sense magnetic fields was dismissed as biological heresy, yet today it stands at the cutting edge of quantum engineering. This chapter bridges the historical struggle for scientific acceptance with the futuristic pursuit of bio-inspired sensors, revealing how the secrets of migrating birds are helping engineers design navigation systems that never go dark.*

## 9.1 The Long Road from Heresy

There is a particular kind of scientific claim that invites ridicule before it invites curiosity: the claim that an animal can sense something humans cannot. To many researchers in the mid-twentieth century, "a magnetic sense" sounded less like physiology and more like the language of stage mentalists. Magnetism was the force behind compasses, motors, and the new world of electronics; it belonged in workshops and physics departments, not in the messy, temperature-wobbly realm of animal behavior. When early biologists suggested that birds might carry a built-in compass, the idea arrived trailing cultural baggage: dowsing rods, mystical "lines" in the landscape, and a century of seductive but sloppy anecdotes.

It did not help that magnetoreception, if it existed, would be *quiet*. The Earth's magnetic field is weak (tens of microtesla), and it does not announce itself with an obvious sensory "feeling" the way light, sound, or odor does. A pigeon returning home cannot tell you which cue it used. A robin in a cage cannot explain why it scratches in a particular direction at dusk. The only way to make the invisible visible is through carefully designed experiments that translate an internal preference into a measurable behavior. That translation—behavior as an instrument—was precisely what made skeptics uneasy. Behavior can be misread, confounded, nudged by subtle cues from the experimenter, or warped by stress. The history of magnetoreception is, in large part, a history of learning how to ask a delicate question without accidentally answering it yourself.

The modern story often begins with a disarmingly simple setup: a migratory songbird placed in a circular cage lined with scratch paper. At migratory seasons, many birds become restless at night, hopping and fluttering in the direction they "want" to fly. This is not flight, of course; it is intent rendered as scuff marks. In the 1960s, Wolfgang and Roswitha Wiltschko, working in Germany, used such orientation cages (and related designs) to test whether European robins would change their preferred direction if the surrounding magnetic field was altered. The crucial tool was not exotic. It was a pair of coils—Helmholtz coils—capable of creating a uniform magnetic field across the cage. With them, an experimenter could rotate the direction of magnetic north without changing the room, the stars outside, or the smell of the air. If the bird's scratches rotated with the field, then something inside the bird was tracking magnetism.

One date has become almost talismanic among those who tell this history: 12 October 1963, associated with a turning point in the Wiltschkos' work. The importance of the moment is not that a single night "proved" magnetoreception; science rarely offers such theatrical clarity. Its importance lies in what the night represents: the shift from vague suspicion to a reproducible behavioral signature, under controls tight enough that critics could not easily dismiss the result as coincidence. Birds exposed to rotated magnetic fields altered their orientation accordingly. The compass, whatever it was, seemed real.

Yet even this kind of result can be attacked from multiple angles. Perhaps the coils produced heat or vibration. Perhaps they emitted a faint hum the birds could hear. Perhaps they influenced the distribution of ions in the air. Perhaps the birds were reacting to the experimenter's handling, or to a pattern of light reflecting off metal. Good skepticism is not simply disbelief; it is the habit of imagining alternative explanations and insisting that an experiment rule them out. The problem in the early years was that magnetoreception demanded an unusually high standard of skepticism because the claim was unusually easy to contaminate. If you are trying to show that a robin is using a magnetic compass, you must convince others that it is *not* using a visual landmark, *not* using the sun's position, *not* reading polarized light at twilight, *not* smelling a draft, and *not* responding to some artifact of your apparatus. You must also convince them that the behavior you are measuring—scratching, hopping, wing-fluttering—is a reliable proxy for a directional choice rather than a stress response.

The early pioneers learned, sometimes painfully, that birds can produce false negatives just as readily as false positives. If an animal is frightened, dehydrated, or otherwise physiologically "off," it may stop expressing the behavior that makes it measurable. That does not mean the sensory system has vanished; it means the experimental window has closed. Magnetoreception research became notorious for this fragility. Some labs reported crisp magnetic effects; others failed to replicate them and concluded the entire phenomenon was illusory. In hindsight, part of the disagreement likely came from details that seemed trivial until they proved decisive: the intensity of the applied field, the stability of the cage environment over days, and the time allowed for acclimation. A bird taken from the outdoors and dropped immediately into an unfamiliar cage inside a heavily controlled room is not merely a "bird in a room." It is an animal abruptly stripped of its spatial context, routine cues, and safety signals. One can imagine a skeptical reviewer reading a report of disoriented scratching and thinking: why assume a magnetic explanation when stress is enough?

In some ways, the controversy echoed an older pattern in biology: the slow acceptance of senses that are not part of human experience. Before microscopes made microbes unavoidable, "germs" were conjecture. Before electrical recording technologies, the notion of the brain as an electrical organ was speculative. Magnetoreception was harder still because even its basic physics felt like a trap. The energies involved in the Earth's field are minuscule compared with thermal noise. How could a warm, wet body register them with any reliability? For decades, critics returned to that intuition: even if birds *behaved* as if guided by magnetism, the mechanism seemed implausible. A phenomenon without a mechanism is not automatically false, but it is socially fragile. It sits in a limbo where every failed replication looks like a disproof and every successful one looks like a fluke.

Mechanisms did exist as hypotheses, and their plurality created another kind of skepticism. If ten plausible mechanisms are proposed, a critic can always say: you are not seeing magnetism; you are seeing an uncontrolled side-effect, because magnetism cannot possibly act the way you claim. In the magnetoreception case, the candidate mechanisms ranged from iron-based particles (biogenic magnetite) acting like microscopic compass needles, to electrical induction effects, to a more counterintuitive idea that arrived from the borderlands between chemistry and quantum physics: the radical-pair mechanism. The very diversity of explanations was a sign of uncertainty, and uncertainty invites dismissal.

One can feel the intellectual mood of the time in the language scientists used. "Extraordinary claims require extraordinary evidence," the adage goes, and magnetoreception was treated as extraordinary. Papers were scrutinized for any hint of circular reasoning. Were the birds truly oriented, or were the researchers "choosing" an angle that fit the expectation? Were the statistics appropriate for circular data? Were the trials blind? Was the equipment calibrated? A few questionable claims in neighboring areas of animal behavior—claims about "unknown forces" or "telepathic" influences—made the atmosphere more combustible. Magnetoreception researchers were forced to police not only their own experiments but the reputation of the entire idea.

A telling episode, repeated in different forms across labs, involved attempts to replicate magnetic orientation using field strengths that were simply wrong. A naive approach might assume that if birds can sense the Earth's field, then stronger fields should produce clearer responses. But biology is not always monotonic. Overstimulate a receptor system and it can saturate, distort, or shut down; apply the wrong geometry and you can scramble what the animal actually uses. The Earth's field is not merely a direction; it has an inclination (the field lines tilt downward in the Northern Hemisphere and upward in the Southern), and it has an intensity that varies with latitude. If the biological compass is tuned to certain ranges or certain components, then a lab field that is "too strong" or "too simplified" might be as uninformative as a blinding spotlight is to a dark-adapted eye. A skeptical team could genuinely believe it had performed a decisive test, only to discover later that its test was like trying to assess hearing by blasting white noise at painful volume.

The gradual hardening of the field—its transition from "maybe" to "real enough to build on"—came from a style of evidence that is underappreciated outside science: converging constraints. The question was never, "Can we get a bird to do something strange in a magnetic field?" The question was, "Can we get a bird to do something *predictably constrained* by the rules of geomagnetism, across conditions, in a way that matches what it would need in the wild?" When researchers showed, for example, that certain birds use an inclination compass—sensitive to the angle of the field lines rather than to magnetic polarity—the result was both bewildering and reassuring. Bewildering, because it implied an unfamiliar biological design (a compass that distinguishes "poleward" from "equatorward" rather than "north" from "south"). Reassuring, because it produced a distinctive behavioral signature that was difficult to fake with simpler explanations. A bird that reverses its orientation when the inclination is inverted, even when "north" and "south" are kept constant, is not responding to a draft or a room landmark. It is responding to something abstract, something geometric about the field itself.

This is where the heresy began to look like a discipline. The best experiments did not merely rotate the magnetic field and watch the bird follow; they changed one parameter at a time and asked whether the animal's behavior obeyed the internal logic of a compass. They separated "map" information (where am I?) from "compass" information (which way is which?), they distinguished intensity cues from directional cues, and they measured what happened when cues were put in conflict. Each refinement was also a reply to a critic: no, it is not vibration, because vibration did not change with inclination; no, it is not a visual artifact, because the effect disappears under certain lighting conditions but returns under others; no, it is not random scratching, because the directional distribution tightens in ways consistent with an internal heading.

The psychological lesson was as important as the technical one. Early magnetoreception work asked scientists to accept that an animal's sensory world—the *Umwelt*, as biologists sometimes call it—contains dimensions we do not share. Humans live inside magnetism in a practical sense; our technologies hum with it. But we do not *perceive* it. That makes it easy to underestimate how richly it might be represented in another brain. Once you take that possibility seriously, a great deal of animal behavior reads differently. A migrating bird is not merely "following instinct." It is sampling a planet-sized field that changes predictably with geography, learning patterns, and combining those patterns with wind, stars, odors, and landmarks. Magnetism is not the whole story, but it can be a stable backbone when other cues fail.

Even so, the field never got the comfort of a single "smoking gun" organ that silenced doubts. Instead, it endured a long middle period in which behavioral evidence was strong but receptor evidence was contentious. Iron-rich particles were reported in bird beaks, but skeptics argued they could be ordinary immune cells filled with iron storage compounds rather than specialized sensors. Light-dependent magnetic effects suggested a retinal mechanism, but translating spin chemistry into nerve impulses seemed like a conceptual leap. The situation was almost perversely educational: it forced researchers to separate *what animals can do* from *how they do it*. Many scientific disputes collapse because people conflate these questions. Magnetoreception endured because the best work kept them apart. An animal can demonstrably use a magnetic cue even if the molecular details remain unsettled. That is not a weakness; it is a normal state of affairs in biology, where function is often mapped long before mechanism is nailed down.

There is another, quieter reason the heresy lingered: magnetoreception research required collaboration across cultures of expertise that did not naturally mix. Behavioral biologists had to become competent with coils, shielding, and field measurements. Physicists had to take seriously the variability of animals and the subtleties of experimental design. Chemists had to entertain the idea that reaction yields could be "read out" by neurons. Each group carried its own instincts about what counts as convincing. A physicist's idea of a clean control is not always a biologist's, and vice versa. When such communities collide, skepticism can look like hostility when it is actually a mismatch of standards. The pioneers were often forced to become bilingual: fluent in animal care and in electromagnetic instrumentation, comfortable discussing both migratory restlessness and vector components of the geomagnetic field.

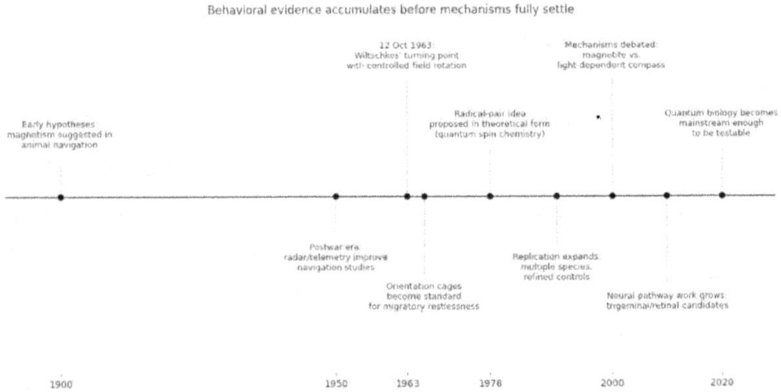


With time, the tone of skepticism shifted. The blunt dismissal—"this is paranormal"—became rarer, replaced by a more productive kind of doubt: *Which* magnetic cues are being used, *when*, and *through what receptors*? That shift matters. It marks the moment a field stops defending its right to exist and starts arguing about details, which is the ordinary music of science.

Yet the old skepticism left a permanent imprint on the field's habits. Magnetoreception researchers tend to be unusually careful about artifacts, unusually alert to hidden electromagnetic noise in laboratory buildings, unusually concerned with blinding and statistics. They learned to respect the ease with which a magnetic effect can be manufactured by accident—by a steel screw in the wrong place, by an electrical cable carrying current behind a wall, by a poorly characterized radiofrequency environment. Many scientists who work on magnetoreception today can tell stories of chasing an effect for weeks only to discover it was caused by a piece of equipment that had nothing to do with the experiment, a reminder that an animal's "compass" is not the only thing in the room that responds to fields.

A final irony is that the heresy was never only about birds. It was about what kinds of explanations biology is allowed to use. The fear was not merely that a magnetic sense might be wrong, but that if it were right it would force biology to borrow concepts that felt alien: spin states, reaction kinetics sensitive to tiny fields, sensory transduction at the edge of physical detectability. That borrowing can feel like a loss of autonomy, as if biology is conceding that its mysteries can be reduced to physics. The better view is the opposite. Biology has always been physics in a crowded room: the same laws, but expressed through architectures that have been tested by evolution rather than by human design. If a robin can treat the Earth's field as usable information, it is not because the field is strong. It is because the animal has found, through countless generations, a way to turn a whisper of physics into a decision about direction.

## 9.2 Engineering the Sixth Sense

The magnetic sense has an odd status in the human imagination. It sounds like a superpower, yet it solves a problem engineers recognize immediately: how to keep your bearings when your usual signals disappear. A bird crossing a featureless sea does not have road signs.

A turtle in mid-ocean cannot glance at a skyline. Even a human hiker, in fog on an unfamiliar moor, learns the uncomfortable truth that vision is not the same thing as orientation. Modern machines have inherited an even sharper version of that vulnerability. Give a robot a clear sky and it can be astonishingly competent; take away satellite navigation and it can become, in an instant, a very expensive amnesiac.

Satellite navigation (GNSS, including GPS) works by measuring how long radio signals take to arrive from multiple satellites. The method is elegant, but it rests on a fragile premise: that those radio signals are available, unspoofed, and strong enough to be trusted. In cities they bounce off glass and concrete; underground and underwater they vanish; in conflict zones they may be jammed or imitated. Engineers have known for decades that any system that collapses when one sensor fails is not a system, but a single point of failure with accessories.

It is tempting to imagine that animals avoid this trap because they possess one magic organ, a built-in compass that is always correct. The reality is more interesting, and more useful: animal navigation behaves like *sensor fusion*—the art of combining multiple imperfect cues into a decision that is good enough to keep moving. The magnetic field is not a replacement for other information; it is a stabilizer, a quiet reference signal that keeps orientation from drifting when everything else becomes unreliable.

Roboticists and sensor engineers have been circling this idea for years, especially in domains where GPS is absent by physics rather than policy. A deep-sea vehicle cannot see the sky. A mining robot cannot climb to the surface every few minutes to get a fix. A rescue drone in a smoke-filled building cannot count on cameras. Each of these settings pushes designers toward the same question: what can play the role of the animal's hidden compass?

One answer is direct: measure the Earth's magnetic field and treat it as a heading reference, just as a phone compass does. That works sometimes, but it runs into a trap that animals also face. The magnetic field is not perfectly smooth. It is distorted by local geology and by human-made structures: steel beams, power lines, vehicles, reinforced concrete. A robot that tries to navigate purely by magnetometer can find itself tugged off course by a hidden staircase or a pipeline like a compass needle near a ship's hull. In practice, the magnetometer becomes a sensor that must be *managed*, not merely read.

A second answer is subtler, and more biomimetic: use the magnetic field not only as a direction, but as a *map*. The field's intensity and inclination vary across the planet in broad, predictable gradients, and local anomalies create distinctive signatures. To an engineer, that begins to look less like a compass and more like terrain. A robot can compare the magnetic pattern it measures to a stored magnetic map and infer where it is, much as a person might recognize a neighborhood by its street layout rather than by a single landmark. This approach has been explored in navigation research under many names—geomagnetic navigation, magnetic anomaly navigation, map-matching—and it has a deep historical resonance.

During the Cold War, submarines faced a problem that sounds modern: they needed to navigate without emitting signals that could be detected. Radio fixes were risky. Surfacing was dangerous. The solution was a mixture of inertial navigation (integrating acceleration and rotation to estimate position) and occasional updates from other cues when safe. Inertial systems drift; tiny errors accumulate into large ones. So navigators sought environmental references that were passive and persistent: gravity anomalies, bathymetry, and magnetic anomalies among them. Long before anyone spoke about "biomimicry," engineers were already learning a lesson that animals embody: when you cannot trust broadcasts from the sky, you look for structure in the planet itself.

That lesson has become newly urgent for robotics. Consider a warehouse robot. It drives between tall metal shelves, under lights, near motors, and beside forklifts—a chaos of electromagnetic clutter. A naive magnetometer heading will be untrustworthy. Yet the clutter is also *stable*. The steel in the building does not rearrange itself every hour. The motors have characteristic patterns. If the robot records magnetic signatures as it moves, it can begin to treat the distortions not as noise to be eliminated but as features to be recognized. This is an engineer's version of turning a weakness into a landmark. It also mirrors what biologists suspect many animals do on smaller scales: not merely detecting "north," but exploiting regularities in the magnetic environment as part of a larger navigational toolkit.

Of course, the Earth's magnetic field is faint. The deeper puzzle is not why engineers want magnetic navigation, but how living systems extract useful information from such a weak signal while surrounded by thermal agitation and biological messiness. That puzzle has acted as a kind of intellectual engine for new sensor ideas, particularly those inspired by the radical-pair mechanism associated with cryptochrome.

A radical pair is formed when light excites a molecule and an electron transfer creates two molecules (or two parts of a molecule) each carrying an unpaired electron. These unpaired electrons have a quantum property called *spin*. The spins can be correlated in ways that influence how the chemical reaction proceeds. External magnetic fields, even weak ones, can slightly alter the interconversion between different spin states, which in turn can bias the reaction's outcome. It is a chain of translation: from a physical field to a microscopic spin dynamic to a chemical yield difference that a cell might, in principle, detect.

For a non-specialist, the provocative part is not the word "quantum." It is the idea that a chemical reaction can act like a sensor. Engineering has tended to separate sensing from chemistry: sensors measure, chemistry reacts. Radical-pair thinking blurs the boundary. The sensor is a reaction network whose products differ depending on the magnetic environment. In a lab, a photochemical system can be illuminated continuously, producing a steady stream of radical pairs; small biases in reaction yield, accumulated over time, can become a signal that is large enough to matter. That accumulation is what biologists call amplification and engineers call integration: you let a tiny effect build until it is no longer tiny.

The attraction for engineers is obvious. If biology can exploit reaction kinetics to make Earth-strength magnetic effects legible, perhaps devices can do something similar. Not necessarily by copying the exact protein, but by copying the *architecture*: *(1)* a light-driven element that reliably produces spin-correlated radicals; *(2)* a chemical stage that converts slight differences into larger differences; *(3)* a readout stage that converts chemistry into an electrical signal; and *(4)* an algorithm that turns that signal into a heading estimate.

There is an anecdote that captures the shift in mindset this demands. In the early days of electronic navigation, a "compass" meant a needle, a mechanical element that physically aligned with a field. When engineers first heard serious proposals for a *chemical* compass, many reacted the way early biologists reacted to magnetoreception itself: with polite disbelief. A compass that does not point, a sensor whose internal state must be inferred statistically, sounded like a category mistake. Yet modern computing has made a new kind of compass practical: one that does not produce a single definitive arrow, but a probability distribution updated continuously as new measurements arrive. This is, in effect, what animals seem to do. A bird does not need a needle in its head; it needs a bias in its neural computations that is consistent enough to guide a migratory decision.

To make that parallel concrete, it helps to lay biological and engineered components side by side. The comparison is not exact—proteins are not circuit boards—but the functional roles can be remarkably similar.

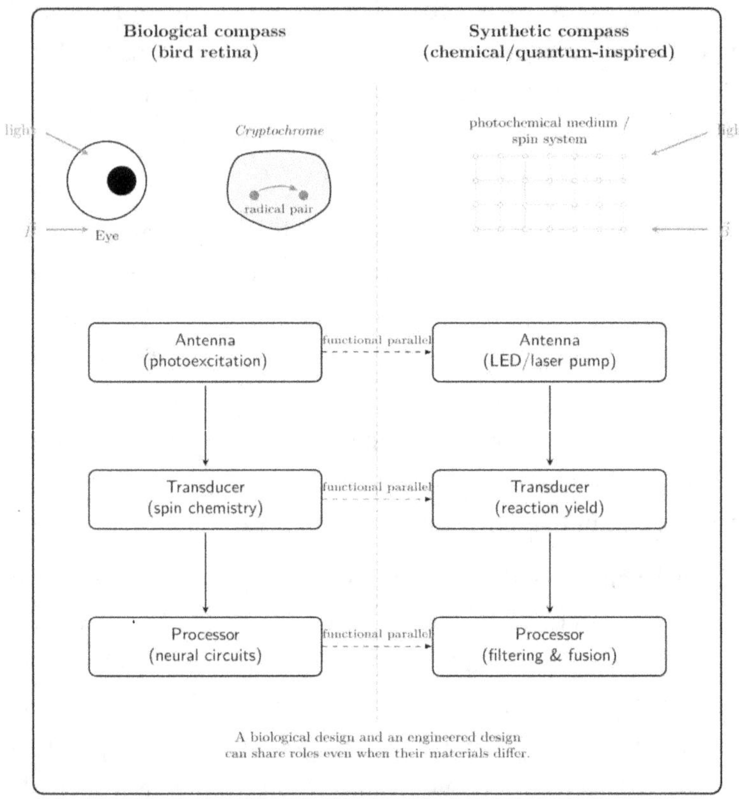


A biological design and an engineered design can share roles even when their materials differ.

The point of this diagram is not to claim that a robot can be built by swapping proteins for printed circuit boards. It is to emphasize a pattern: both systems need a way to couple magnetism to an internal variable, a way to amplify weak effects, and a way to interpret the result amid noise.

Some engineering efforts stay close to conventional hardware while borrowing the biological logic. A magnetometer can be made far more sensitive than a phone compass. Optical magnetometers and diamond-based magnetometers (notably those using nitrogen-vacancy centers) can measure tiny fields with impressive precision. These devices are not cryptochrome, but they share an important property with radical-pair thinking: they treat magnetism as something that can be read through a state transition and then detected optically or electrically. In a laboratory, such sensors can be extremely sensitive. The challenge is taking them out of the lab and into the environments where GPS fails: vibrations, temperature swings, power constraints, and electromagnetic interference.

Even with excellent sensors, navigation is not simply a measurement problem. It is an inference problem. A robot rarely asks, "What is the magnetic field?" It asks, "Given what I measured, where am I and which way am I facing?" The answer depends on context. A magnetometer reading on its own is ambiguous; the same heading can occur in many places. The engineered solution is to combine magnetic information with inertial measurement units (accelerometers and gyroscopes), wheel odometry, cameras when available, sonar or lidar when available, and sometimes cooperative measurements shared among multiple robots. The mathematical frameworks behind this are often filters—extended Kalman filters, unscented Kalman filters, particle filters—names that sound forbidding but describe a simple idea: keep track of your best guess, keep track of how uncertain that guess is, and update both as new evidence arrives.

Here the biological analogy becomes emotionally vivid. A migrating bird also lives with uncertainty. Wind pushes it off course. Clouds erase the stars. The magnetic field varies and sometimes becomes locally distorted. Yet the animal continues. It does not demand perfect information; it demands sufficient information to keep making progress. Engineers have learned the same pragmatic stance. In a GNSS-denied environment, the goal is not to recreate the crisp certainty of GPS. The goal is resiliency: graceful degradation instead of catastrophic failure.

A concrete example makes the difference clear. Imagine an autonomous underwater vehicle tasked with surveying a pipeline. Underwater, GPS is unavailable. The vehicle relies on inertial navigation, but drift accumulates. If the vehicle carries a magnetometer and has access to a magnetic anomaly map of the seafloor region (built from previous surveys or from public geophysical data at suitable resolution), it can periodically compare its measured magnetic signature to the map and correct its position estimate. The correction is not constant; it depends on how distinctive the magnetic pattern is. Over a smooth region, the map provides little information. Over an area with strong, unique anomalies, the map becomes a set of fingerprints. This looks less like a compass and more like recognizing a melody: a sequence of rises and falls that is hard to confuse with another.

The same philosophy can be applied in buildings, where robots operate among steel and power. The robot can first learn a magnetic map during a training run, storing intensity and direction anomalies along with its estimated path. Later, when GPS is absent and vision is degraded, the robot can localize itself by matching current magnetic measurements to its learned magnetic landscape. Some systems go further and treat magnetic disturbances as hazards to be detected and handled explicitly: if the magnetic field becomes inconsistent with past experience, the robot can downweight the magnetometer and lean on other sensors until it escapes the disturbance. That is not far from what a cautious animal might do: a cue is valuable until it becomes unreliable, and reliability can be estimated.

Biomimicry enters not only through sensor choice but through strategy. Animals rarely bet everything on a single cue; they use magnetic information as a scaffold for other decisions. Engineers are beginning to build robots that do something similar: a magnetic heading estimate can stabilize a camera-based tracker; a magnetic map match can reset an inertial system; a learned model can predict what "normal" magnetic measurements look like along a route, flagging anomalies that might correspond to an unexpected object or a changed environment. The magnetic sense, in this reading, is not a gadget. It is a way of organizing uncertainty.

The radical-pair idea continues to exert a particular fascination because it suggests that highly sensitive magnetic sensing might be possible with low power, small size, and materials very different from traditional magnetometers. It also suggests a novel control knob: light. A light-driven sensor can, in principle, be tuned by illumination wavelength, intensity, and modulation. That is appealing in robotics, where controllable excitation can improve signal extraction. Some experimental work on cryptochrome-related spin chemistry has even shown that radiofrequency fields can perturb magnetic sensitivity in ways that reveal the underlying spin dynamics, hinting at the possibility of sensors that are not merely passive but actively interrogated. For an engineer, active interrogation is a familiar move: radar is active sensing; lidar is active sensing. A chemical compass that benefits from controlled illumination begins to look less mystical and more like an instrument.

A sober view is also necessary. The leap from "a protein may respond to Earth-strength fields" to "a robot can navigate using a cryptochrome-like device" is enormous. Biology has the advantage of exquisite molecular organization and deeply integrated amplification pathways. An engineered version must replace those with manufacturable components and robust readouts. Yet even if a direct cryptochrome sensor never becomes a practical navigation module, the *conceptual* gift remains: it enlarges the engineer's imagination about what a compass can be. It need not be a needle. It can be a probability distribution derived from weak biases, accumulated and filtered until a direction emerges.

There is a final twist that brings the story back to resiliency. The Earth's magnetic field is not owned by anyone. It cannot be switched off in a city during a blackout. It does not require a subscription. It does not depend on satellites that can fail, or signals that can be jammed. It is a planetary background condition, imperfect and locally distorted but stubbornly present. Animals evolved in that background and learned to treat it as information. When engineers borrow that habit of mind—treating the planet itself as a reference rather than treating navigation as a service provided from above—they are not building a magical new sense so much as recovering an old idea: orientation can be anchored in the world, not just in the network.

## 9.3 The Needle in the Haystack

It is possible to know that a sense exists, to measure what it can do, to chart the situations in which it fails, and still not know where it begins. That is the peculiar state of magnetoreception. Behavioral experiments can show that an animal responds to controlled changes in a magnetic field. Navigation studies can show that magnetic cues matter in the wild. Physics can offer mechanisms that are plausible at Earth-strength fields. Yet the most satisfying kind of answer—the one that lets you point to a cell and say, *this is where magnetism enters the nervous system*—remains frustratingly rare.

The difficulty is not simply that the relevant cells are small. Many sensory cells are small. The difficulty is that magnetoreception is, almost by definition, a sense for a subtle, omnipresent stimulus. You cannot easily "turn off" the Earth's field in the way you can turn off a light or silence a sound. You cannot easily deliver a neat, localized magnetic "puff" that stimulates only one patch of tissue. A magnetic field passes through skull and skin without caring about boundaries. Even worse, living tissue contains iron, conducts electricity, and is full of moving charges; the very act of measuring can create artifacts. The haystack is not only large. It is conductive, warm, and full of needles that are not the needle you are looking for.

A useful starting point is to name what is being sought. A *primary receptor* is the first biological element that converts a physical stimulus into a biological signal: a change in membrane voltage, a chemical messenger, a change in firing rate, something the nervous system can use. In vision, this role belongs to photoreceptor cells in the retina, which contain light-sensitive pigments. In hearing, it belongs to hair cells in the inner ear, which convert mechanical vibration into electrical changes. A magnetoreceptor cell would play the analogous role for magnetic stimuli. Once such a cell is identified and shown to be necessary and sufficient for magnetic sensing in a living animal, the field changes. Arguments about whether the sense is real become historical curiosities, and the focus shifts to wiring, computation, and evolution.

The quest has been complicated by the likelihood that there is not one magnetoreception, but several. Behavior suggests at least two different kinds of magnetic information are used by birds: a compass-like sense that gives directional information, and a map-like sense that may provide intensity-related information, possibly used in navigation and location. These could, in principle, rely on different receptors and different neural pathways. Even within "compass" behavior, light dependence points toward the eye in some species and contexts, while other experiments implicate structures associated with the beak and the trigeminal nerve. The puzzle begins to resemble trying to locate the "seat of balance" in the body without knowing about inner ears: you might find several contributing systems before you find the one you were expecting.

The magnetite hypothesis seems, at first glance, like the easier target. Magnetite is an iron oxide that is naturally magnetic; it behaves, in miniature, like the material in a compass needle. Many organisms produce magnetite for other purposes, and magnetotactic bacteria famously use chains of magnetite crystals to align with the Earth's field. If birds or other animals had magnetite particles anchored in a cell, mechanical torque from the field could, in principle, open ion channels or deform membranes, producing an electrical signal. This is a straightforward story by biological standards: a physical force becomes a cellular signal via mechanosensation.

Reports of iron-rich structures in bird beaks once looked like the beginning of closure. The beak is richly innervated, and the trigeminal nerve is a major pathway for facial sensation. Some behavioral and physiological evidence suggested that changing magnetic intensity (not just direction) could influence trigeminal pathways. If true, it offered a satisfying division of labor: the eye for compass direction, the beak for map intensity. For a while, it was possible to imagine that a careful histological search would reveal magnetite-bearing receptor cells, perhaps arranged in clusters, perhaps aligned in a meaningful geometry, and that the argument would end in the way many arguments in sensory biology end: with a micrograph and a tracing of the nerve.

Instead, the beak story became a cautionary tale about how hard it is to interpret "magnetic-looking" anatomy. Iron accumulates in many cell types for ordinary physiological reasons. Immune cells called macrophages can contain iron-rich granules as part of their role in recycling and storage. A cluster of iron-containing cells is not automatically a cluster of sensors; it may be a cluster of housekeeping. Some influential critiques argued that structures previously labeled as magnetoreceptors could be explained as macrophages rather than specialized sensory endings. The controversy was not merely semantic. If the "receptors" are macrophages, then the most direct mechanistic story collapses, and the field must ask whether the trigeminal evidence points to a different receptor type entirely, perhaps located elsewhere, perhaps more dispersed than expected.

At this point, a dramatic kind of skepticism returned, not about whether birds sense magnetism, but about whether the most photogenic candidate receptors were ever what they seemed. It was a painful moment for the community because it replayed an old dynamic: an appealing anatomical picture draws attention, and then a more mundane explanation threatens to erase it. Yet it also clarified something important. A receptor is not defined by its appearance, but by its causal role. You can stain iron until it glows and still learn nothing about sensation if you cannot show that the stained structure is *necessary* for the behavior and that it *responds* to magnetic stimuli in a way that can drive neural activity.

Evidence for trigeminal involvement did not vanish with the anatomical dispute. Experiments in pigeons have reported activation of trigeminal brainstem nuclei when magnetic fields change in certain ways, with effects reduced or abolished under zero-field conditions or after cutting the ophthalmic branch of the trigeminal nerve. Such results are hard to dismiss as mere coincidence. They suggest that, whatever the receptor is, information about magnetic changes can reach the brain via trigeminal pathways. The receptor might not be a tidy, compass-needle-like structure in the beak; it might be a more distributed system, or a structure in nearby tissues, or even a receptor that relies on properties of iron without being a classic magnetite "needle."

If the magnetite search has been bedeviled by biological ambiguity, the radical-pair search has been bedeviled by conceptual delicacy.

The radical-pair mechanism is attractive because it naturally explains why magnetic compass behavior in some birds depends on light and can be disrupted by certain radiofrequency fields. It also places the sensor in the retina, a tissue already built for detecting and processing subtle physical signals. But a plausible physical mechanism is not the same thing as a demonstrated receptor cell. The retina contains many cell types, many photopigments, and many molecular pathways that respond to light. Cryptochromes are present in various tissues and play roles in circadian rhythms as well as in photochemistry. Finding cryptochrome in the eye does not, on its own, reveal which cells are doing magnetosensation, what the readout variable is, or how that variable is coupled to neural firing.

Here, the "needle in the haystack" becomes almost literal. Imagine trying to identify the specific photoreceptors responsible for a subtle effect when the retina is already a densely layered sheet of neurons, packed with light-sensitive molecules, constantly adapting to illumination. The magnetic effect, if it exists at the molecular level, may be small: a slight change in reaction yield, a slight shift in the balance between chemical products. The nervous system could amplify that difference, but amplification does not necessarily make the source obvious to an outside observer. It may take place over many synapses, spread over a network, or be encoded in a pattern rather than in a single firing rate.

Theoretical work has sharpened this challenge rather than relieving it. In radical-pair models, magnetosensitivity can be enhanced or suppressed by details that sound, to non-specialists, like technical footnotes: how far apart two radicals are, whether they move relative to one another, how strongly their spins interact through dipolar coupling, whether there is electron hopping along a chain of amino acids, whether the initial spin state carries polarization or coherence. These details matter because Earth's field is weak; there is no margin for sloppy design. A small change in molecular geometry can turn a sensitive compass into an insensitive one.

This is not an abstract concern. Some modeling work has suggested that electron—electron dipolar interactions in a static radical pair can suppress magnetosensitivity, essentially drowning out the effect of the geomagnetic field unless some dynamic process reduces the coupling. Other theoretical studies explore whether motion of a radical partner—diffusing or moving along a constrained path—could intermittently weaken coupling and restore sensitivity, invoking phenomena akin to quantum Zeno-like behavior, where frequent interactions can alter the evolution of a quantum state. There are also proposals involving chiral-induced spin selectivity (CISS), a mechanism by which chiral molecules can preferentially transmit electrons of a given spin. Some recent analyses suggest that CISS may enhance sensitivity in certain cases (for example when it generates spin polarization), but not in others (when it merely generates coherence). The pattern is consistent: there is no universal "quantum trick" that automatically makes a biological compass work. The physics is conditional, and biology must satisfy those conditions in vivo.

For experimentalists, this theoretical richness is both a gift and a burden. It offers testable signatures: light wavelength dependence, sensitivity to oscillating fields, specific angular responses. At the same time, it expands the parameter space. If you cannot find the receptor cell, perhaps it is because you are looking in the wrong tissue; or in the right tissue but at the wrong developmental stage; or under the wrong lighting; or with an assay that misses the relevant readout. When a phenomenon depends on a chain of delicate conditions, a null result is hard to interpret. It might mean "no magnetoreception," or it might mean "magnetoreception is there, but your setup broke one link in the chain."

A historical anecdote from sensory biology helps to put the frustration in proportion. Long before the molecular identity of photopigments was known, scientists could demonstrate that animals responded to light and even infer some properties of color vision. Yet identifying the exact molecules and cells responsible required the development of stains, electrophysiology, and later molecular genetics. For magnetoreception, the technology curve has been steeper because the signal is subtler and the confounds are stranger. Electrophysiology can record neural responses, but a magnetic stimulus is hard to localize. Imaging can show activated brain regions, but activation is not identity; it is the shadow of a computation, not the input stage itself.

The search has therefore become a kind of methodological arms race, and some of the most exciting developments are tools rather than discoveries. Genetic methods can now label specific cell types, knock out candidate proteins, and observe the behavioral consequences. If cryptochrome is required for a magnetic compass, then disrupting cryptochrome in a targeted way should alter magnetic orientation without broadly destroying vision. That is a hard experiment, because cryptochromes can have multiple roles, and the disruption must be precise enough to avoid general sickness or disorientation. But it is the kind of hard that is productive: it turns philosophical debate into a clear causal question.

Microscopy has also advanced in ways that matter here. High-resolution imaging can locate proteins within cells; correlative methods can link molecular labels to ultrastructure. If magnetite-based receptors exist, modern techniques can characterize mineral composition and crystal structure with far more precision than earlier iron stains could provide. The challenge is to do this in vivo or in tissues preserved in a way that does not introduce contamination. Iron is ubiquitous, and magnetite can be created or altered during sample preparation. The line between "what the animal built" and "what the lab produced" is thin.

Even behavioral assays are evolving. The classic orientation cage remains powerful, but newer approaches combine behavioral tracking with controlled manipulation and neural recording. When an animal makes a magnetic-based decision, which neurons change their activity? Which brain regions light up? Can the same magnetic manipulation that changes behavior also change a measurable neural signal? The moment when a magnetic perturbation produces a consistent, time-locked response in a defined neural population is the moment the haystack begins to thin.

One reason this matters beyond academic satisfaction is that biology and engineering have started to lean on each other. Engineers building resilient navigation systems want to know what animals are actually measuring, because it affects what is worth copying. If animals use inclination rather than polarity, a sensor that returns polarity may mislead. If animals treat magnetic information probabilistically and fuse it with other cues, then a robot built around a single "perfect" magnetometer may be less robust than a robot with a merely adequate sensor and a good inference algorithm. Conversely, engineering can offer biology new instruments. Extremely sensitive magnetometers can characterize the electromagnetic environment of an animal experiment with a fidelity that was previously impossible, revealing hidden noise sources that might have confounded results. Techniques from signal processing and control theory can help interpret behavioral variability not as a nuisance but as information about underlying computations.

The holy grail, then, is not simply a cell with a magnetic particle or a protein that forms radical pairs. It is a full causal chain: a physical stimulus producing a change in a defined molecular system, producing a change in a defined cellular signal, producing a change in a defined neural pathway, producing a change in a measurable behavior. That is a demanding standard, but it is the standard that turns magnetoreception from an intriguing capability into a tractable sensory system.

For all the frustration, there is something quietly inspiring in how stubborn the problem has been. The Earth's field is a faint thread woven through every environment, and animals appear to have learned to tug on that thread without snapping it. If the receptor cells are sparse, that may be because they do not need to be many; a small bias, amplified and combined with other cues, can steer an entire migration. If the mechanism is delicate, that may be because it operates at the edge of what physics allows in warm biology, a place where evolution can be surprisingly inventive.

In the lab, the haystack sometimes looks endless: tissues sliced into thin sections, stained and restained, candidate structures appearing and dissolving under new interpretations, models that work on paper but stumble on realistic constraints. Yet the search is also gradually acquiring the satisfying texture of other success stories in sensory biology: hypotheses sharpening into predictions, predictions forcing better experiments, and better experiments revealing that what looked like magic was a specific piece of anatomy doing a specific piece of work. Somewhere in an eye, a beak, or a nervous system not yet examined with the right combination of patience and tools, a microscopic transducer is quietly converting a planetary field into a biological fact.

www.ingramcontent.com/pod-product-compliance
Lightning Source LLC
LaVergne TN
LVHW022206200926
848481LV00006B/1864
* 9 7 9 8 8 9 6 6 5 2 7 7 9 *